ROTARY DRILLING SERIES

Drilling Fluids

Unit II, Lesson 2
First Edition

▼
▼
▼

by Kate Van Dyke

Published by

PETROLEUM EXTENSION SERVICE
The University of Texas at Austin
Division of Continuing & Innovative Education
Austin, Texas

originaly produced by

INTERNATIONAL ASSOCIATION OF DRILLING CONTRACTORS
Houston, Texas

2000

Library of Congress Cataloging-in-Publication Data

Van Dyke, Kate, 1951—
Drilling fluids / by Kate Van Dyke. — 1st ed.
p. cm. — (Rotary drilling series ; unit II, lesson 2)
ISBN 0-88698-189-1
1. Drilling muds. I. Series

TN871.27.V36 2000
622'.3381—dc21 00-028634
CIP

First Edition published 2000.
Fifth Impression 2012
Printed in the United States of America

Catalog no. 2.202101
ISBN 0-88698-189-1

No state tax funds were used to publish this book.
The University of Texas at Austin is an equal opportunity employer.

Contents

Figures

▼
▼
▼

Tables

Foreword

▼
▼
▼

For many years, the Rotary Drilling Series has oriented new personnel and further assisted experienced hands in the rotary drilling industry. As the industry changes, so must the manuals in this series reflect those changes.

The revisions to both the text and illustrations are extensive. In addition, the layout has been "modernized" to make the information easy to get; the study questions have been rewritten; and each major section has been summarized to provide a handy comprehension check for the reader.

PETEX wishes to thank industry reviewers—and our readers—for invaluable assistance in the revision of the Rotary Drilling Series. We especially wish to thank Tom Thomas, Modular Training Coordinator, Sedco Forex, for his unfailing and extensive help in answering questions, providing technical input, and encouragement.

Drilling fluid is vital to the entire rotary drilling process. Indeed, without a circulating drilling fluid, nobody could successfully drill most wells with the rotary method. What's more, the success or failure of the mud program can largely determine whether the drilling contractor can drill the well to the operator's specifications in a safe and economical way. Although a specially trained mud engineer often bears the responsibility of keeping the right fluid, properly treated to obtain maximum results, in the hole at all times, all rig personnel should have a basic understanding of what goes into a mud program. This book should help.

Although every effort was made to ensure accuracy, this manual is intended to be only a training aid; thus, nothing in it should be construed as approval or disapproval of any specific product or practice.

Ron Baker

Units of Measurement

Throughout the world, two systems of measurement dominate: the English system and the metric system. Today, the United States is almost the only country that employs the English system.

The English system uses the pound as the unit of weight, the foot as the unit of length, and the gallon as the unit of capacity. In the English system, for example, 1 foot equals 12 inches, 1 yard equals 36 inches, and 1 mile equals 5,280 feet or 1,760 yards.

The metric system uses the gram as the unit of weight, the metre as the unit of length, and the litre as the unit of capacity. In the metric system, for example, 1 metre equals 10 decimetres, 100 centimetres, or 1,000 millimetres. A kilometre equals 1,000 metres. The metric system, unlike the English system, uses a base of 10; thus, it is easy to convert from one unit to another. To convert from one unit to another in the English system, you must memorize or look up the values.

In the late 1970s, the Eleventh General Conference on Weights and Measures described and adopted the Système International (SI) d'Unités. Conference participants based the SI system on the metric system and designed it as an international standard of measurement.

The *Rotary Drilling Series* gives both English and SI units. And because the SI system employs the British spelling of many of the terms, the book follows those spelling rules as well. The unit of length, for example, is *metre*, not *meter*. (Note, however, that the unit of weight is *gram*, not *gramme*.)

To aid U.S. readers in making and understanding the conversion to the SI system, we include the following table.

English-Units-to-SI-Units Conversion Factors

Quantity or Property	English Units	Multiply English Units By	To Obtain These SI Units
Length, depth, or height	inches (in.)	25.4	millimetres (mm)
		2.54	centimetres (cm)
	feet (ft)	0.3048	metres (m)
	yards (yd)	0.9144	metres (m)
	miles (mi)	1609.344	metres (m)
		1.61	kilometres (km)
Hole and pipe diameters, bit size	inches (in.)	25.4	millimetres (mm)
Drilling rate	feet per hour (ft/h)	0.3048	metres per hour (m/h)
Weight on bit	pounds (lb)	0.445	decanewtons (dN)
Nozzle size	32nds of an inch	0.8	millimetres (mm)
Volume	barrels (bbl)	0.159	cubic metres (m^3)
		159	litres (L)
	gallons per stroke (gal/stroke)	0.00379	cubic metres per stroke (m^3/stroke)
	ounces (oz)	29.57	millilitres (mL)
	cubic inches ($in.^3$)	16.387	cubic centimetres (cm^3)
	cubic feet (ft^3)	28.3169	litres (L)
		0.0283	cubic metres (m^3)
	quarts (qt)	0.9464	litres (L)
	gallons (gal)	3.7854	litres (L)
	gallons (gal)	0.00379	cubic metres (m^3)
	pounds per barrel (lb/bbl)	2.895	kilograms per cubic metre (kg/m^3)
	barrels per ton (bbl/tn)	0.175	cubic metres per tonne (m^3/t)
Pump output and flow rate	gallons per minute (gpm)	0.00379	cubic metres per minute (m^3/min)
	gallons per hour (gph)	0.00379	cubic metres per hour (m^3/h)
	barrels per stroke (bbl/stroke)	0.159	cubic metres per stroke (m^3/stroke)
	barrels per minute (bbl/min)	0.159	cubic metres per minute (m^3/min)
Pressure	pounds per square inch (psi)	6.895	kilopascals (kPa)
		0.006895	megapascals (MPa)
Temperature	degrees Fahrenheit (°F)	$\frac{°F - 32}{1.8}$	degrees Celsius (°C)
Thermal gradient	1°F per 60 feet	—	1°C per 33 metres
Mass (weight)	ounces (oz)	28.35	grams (g)
	pounds (lb)	453.59	grams (g)
		0.4536	kilograms (kg)
	tons (tn)	0.9072	tonnes (t)
	pounds per foot (lb/ft)	1.488	kilograms per metre (kg/m)
Mud weight	pounds per gallon (ppg)	119.82	kilograms per cubic metre (kg/m^3)
	pounds per cubic foot (lb/ft^3)	16.0	kilograms per cubic metre (kg/m^3)
Pressure gradient	pounds per square inch per foot (psi/ft)	22.621	kilopascals per metre (kPa/m)
Funnel viscosity	seconds per quart (s/qt)	1.057	seconds per litre (s/L)
Yield point	pounds per 100 square feet (lb/100 ft^2)	0.48	pascals (Pa)
Gel strength	pounds per 100 square feet (lb/100 ft^2)	0.48	pascals (Pa)
Filter cake thickness	32nds of an inch	0.8	millimetres (mm)
Power	horsepower (hp)	0.7	kilowatts (kW)
Area	square inches ($in.^2$)	6.45	square centimetres (cm^2)
	square feet (ft^2)	0.0929	square metres (m^2)
	square yards (yd^2)	0.8361	square metres (m^2)
	square miles (mi^2)	2.59	square kilometres (km^2)
	acre (ac)	0.40	hectare (ha)
Drilling line wear	ton-miles (tn•mi)	14.317	megajoules (MJ)
		1.459	tonne-kilometres (t•km)
Torque	foot-pounds (ft•lb)	1.3558	newton metres (N•m)

Introduction

▼
▼
▼

Without a circulating drilling fluid, rotary drilling would be difficult if not impossible in some cases. A fluid is any substance that flows, so drilling fluid may be either liquid, gas, or a mixture of the two. If liquid, drilling fluid may be water, oil, or a combination of water and oil. Operators often put special substances (additives) in these liquids to give them characteristics that make it possible or easier to drill the hole. Most oilfield workers call liquid drilling fluid drilling mud. A gaseous drilling fluid may be (1) dry air or natural gas, (2) air or gas mixed with a special foaming agent, which forms mist or foam or, (3) air or gas mixed with liquid, which is an aerated drilling mud.

Drilling fluids were simple in the early days of rotary drilling; operators usually just used whatever water was available. They dug an open pit in the ground next to the rig and filled it with water. If they wanted to stabilize the hole—that is, keep the hole from caving in where it penetrated soft formations—they stirred up the pit holding the water. Stirring the pits mixed the natural clays in the soil with the water. The solid clay particles plastered the sides of the hole with wall cake. This wall cake often prevented soft formations from caving or sloughing (pronounced sluffing) into the hole. Legend has it that the Hamil brothers, who successfully drilled the Spindletop well in 1901, ran cattle through their water pits to stir up the clay. Whatever they did to make mud, it worked. The solids in the natural clay formed a wall cake on a troublesome formation that enabled the Hamils to successfully drill it. The formation had thwarted several previous attempts when the drillers merely used clear water as a drilling fluid.

Today, drilling fluid choices are complex. The development of special types of fluids and additives to cure all sorts of downhole problems brings its own difficulties. Research and field employees of drilling fluid companies, often called mud engineers, must ask, for example, what additive will best correct a particular drilling problem? How will the mud react to changes in the formation? Will a certain additive interfere with or cancel the effect of another? Will the expense of disposing of a fluid with toxic additives outweigh the benefits of using it?

A great deal of study, analysis, and expense go into a *mud program*, which is the plan for the type and properties of drilling fluid to use. A good mud program ensures that the physical and chemical properties of the fluid are the best ones possible for a particular drilling situation. Miscalculation can result in unnecessary costs in time and money. Although the design and maintenance of a mud program are the responsibility of the mud engineer, all rig personnel will be better equipped to do their jobs if they understand the basics of drilling fluids.

▼
▼
▼

Functions of Drilling Fluids

▼
▼
▼

When drilling with a rotary rig, the rig operator normally circulates some type of fluid. Drilling fluid is necessary because it—

1. cleans the bottom of the hole,
2. transports bit cuttings to the surface,
3. cools and lubricates the bit and drill stem,
4. supports the walls of the wellbore,
5. prevents the entry of formation fluids into the well,
6. transmits hydraulic power to downhole equipment,
7. reveals the presence of oil, gas, or water that may enter the circulating fluid from a formation being drilled, and
8. reveals information about the formation by means of the cuttings the fluid brings up to the surface.

Many kinds of drilling fluids are available, and all perform these functions in their own way. Air and gas have particular advantages and disadvantages compared to drilling muds.

Cleaning the Bottom of the Hole

Regardless of the drilling fluid in use, a bit must have a clean surface on which to work when making hole. For the bit cutters to regrind the chips the cutters have already broken off from the bottom of the hole is wasted effort. If the drilling fluid does not sweep away all or most of the cuttings as the bit makes them, the bit redrills the cuttings, which reduces the rate of penetration (ROP). If the hydraulics are working right—that is, if most of the pressure put on the drilling fluid by a pump or a compressor is expended at the bit—drilling fluid leaves the bit in high-velocity streams, or jets. These high-speed jets of fluid blast the bottom of the hole, creating turbulence that moves the cuttings away from the face of the formation as fast as the bit cuts them (fig. 1). Rig operators carefully size the bit nozzles to ensure that the biggest pressure drop occurs as the drilling fluid leaves the bit.

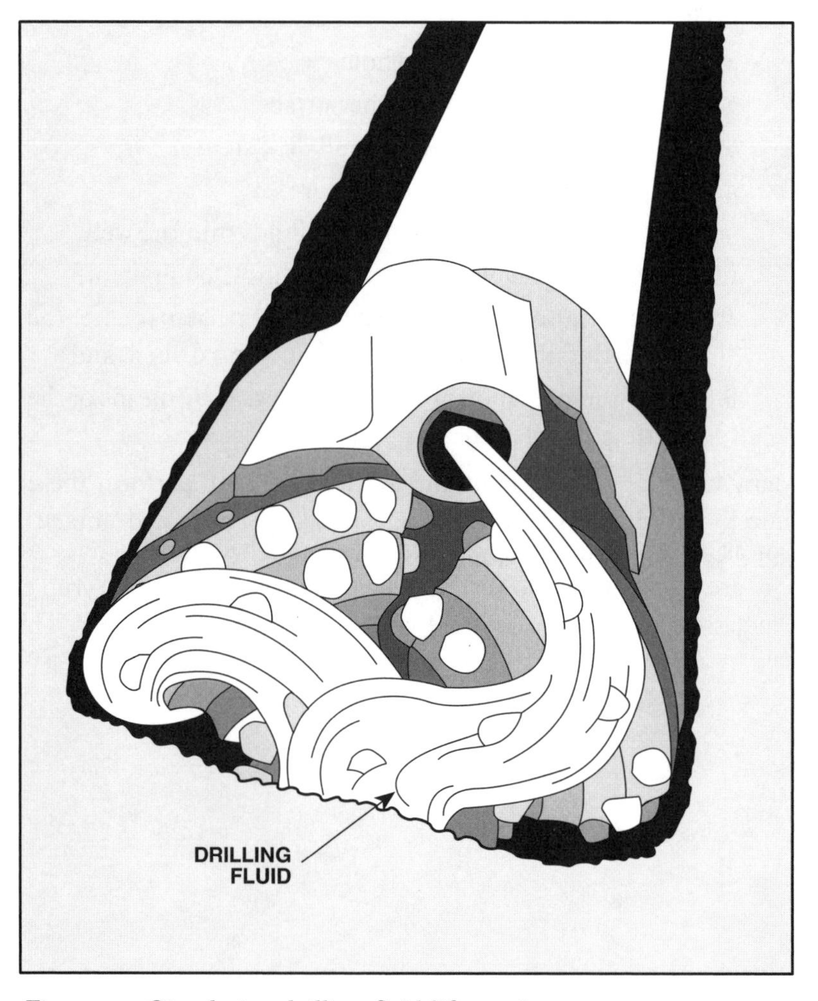

Figure 1. Circulating drilling fluid lifts cuttings.

Transporting Cuttings to the Surface

Drilling fluid carries rock chips, sand, or shale particles from the bottom of the hole as it moves up the annulus. Drilling muds have the additional job of suspending cuttings in the annulus when the driller stops circulating. When crew members make a connection, for example, the driller stops the pumps and circulation ceases until they finish the connection. Drilling fluids such as air and gas cannot suspend cuttings when circulation stops, so the cuttings fall to the bottom. Unlike muds, which a pump recirculates after the mud passes through surface equipment to remove cuttings and other solids, air, gas, and foam normally only go down the hole and up to the surface once. Such gaseous fluids and the solids they carry pass from the hole through a large pipe (the blooey line) and into a waste pit for later disposal.

Annular Velocity

When drilling with air or gas, the annular velocity, or speed at which the fluid flows up the annulus, determines how well cuttings move to the surface. The only way air or gas can move cuttings to the surface is by means of pressure and velocity because a gas has virtually no density to mechanically lift them. Air or gas drilling usually reduces the cuttings to dust by the time they reach the surface because they move so fast and pound against the tool joints and wellbore on their way up. Gas or air circulation requires an annular velocity of about 3,000 feet (ft) or 100 metres (m) per minute (/min).

Transport of cuttings when drilling with a liquid depends on annular velocity, viscosity, and gel strength. For drilling mud, the annular velocity must usually be from 100 to 200 ft/min (30 to 60 m/min) to keep the hole clean. The rate at which cuttings settle, the *slip velocity*, is affected by the viscosity and gel strength of the mud. Viscosity is a measure of a fluid's ability to flow. In general, fluids with high viscosity are thicker (flow slower) than fluids with low viscosity. Syrup or molasses, for example, is more viscous (is thicker) than water. This difference in viscosity can be observed by pouring the two liquids out of their containers. If the temperature of both liquids is the same, the water pours much faster than the molasses because water is less viscous (is thinner). High viscosity drilling muds tend to transport cuttings at lower velocities than low viscosity drilling muds.

Gel strength is a measure of a drilling mud's ability to suspend cuttings in the hole when circulation stops. The higher a mud's gel strength, the more pump pressure it takes to get it moving again after it gels. In general, high-gel strength muds transport cuttings at lower velocities than low-gel strength muds.

Cooling the Bit and Lubricating the Drill Stem

The combination of rotation and weight on bit creates several hundred degrees of heat because of friction. This friction results from the moving parts of the bit and from the bit's cutting elements moving against the formation. For example, the weight on an 8 ½-inch (in.) or 216-millimetre (mm) bit may be 60,000 pounds (lb) or 26,700 decanewtons (dN) of force, about the weight of a railroad freight car. A large-diameter bit may require double that amount of weight. The rotation speed may be as fast as 180 revolutions per minute (rpm) or 3 revolutions per second. Heat is the enemy of metal and especially of the diamonds in a diamond bit. Unless the heat from friction is removed, an expensive bit overheats and quickly wears out. Besides the bit, friction creates heat wherever the drill stem contacts the side of the hole. Fluid circulating through and around the parts of the bit and the drill stem removes the heat.

Air or gas circulation is very efficient for cooling because the air or gas expands as it leaves the bit nozzles, which produces a cooling effect. Perhaps you have noticed that air released from an inflated tire or soccer ball feels cool. This is one reason bits last longer in air or gas drilling.

In general, drilling mud also lubricates the areas where the bit and drill stem contact the hole. While water is not as good as oil as a lubricant, water nevertheless reduces friction between moving surfaces. Of course, if the operator adds oily substances to the drilling fluid, they can reduce friction in the bit bearings and serve as a better lubricant than water where the drill stem contacts the walls of the hole.

Supporting the Walls of the Well

A drilling mud with the proper characteristics can keep a formation from caving, or sloughing, into a well. The weight, or density, of the fluid column in the hole creates downhole pressure. This pressure forces the liquid part (the filtrate) of the mud a short distance into any porous and permeable formations opposite the wellbore (fig. 2). The solid particles in the mud stick to the walls of the hole and form a virtually impermeable sheath (fig. 3). This impermeable wall, or filter, cake plasters the hole wall and stabilizes the formation to keep it from sloughing into the wellbore. Because dry air and gas drilling fluids do not form a wall cake, air or gas drilling is practical only in formations made of hard rock with little tendency to slough.

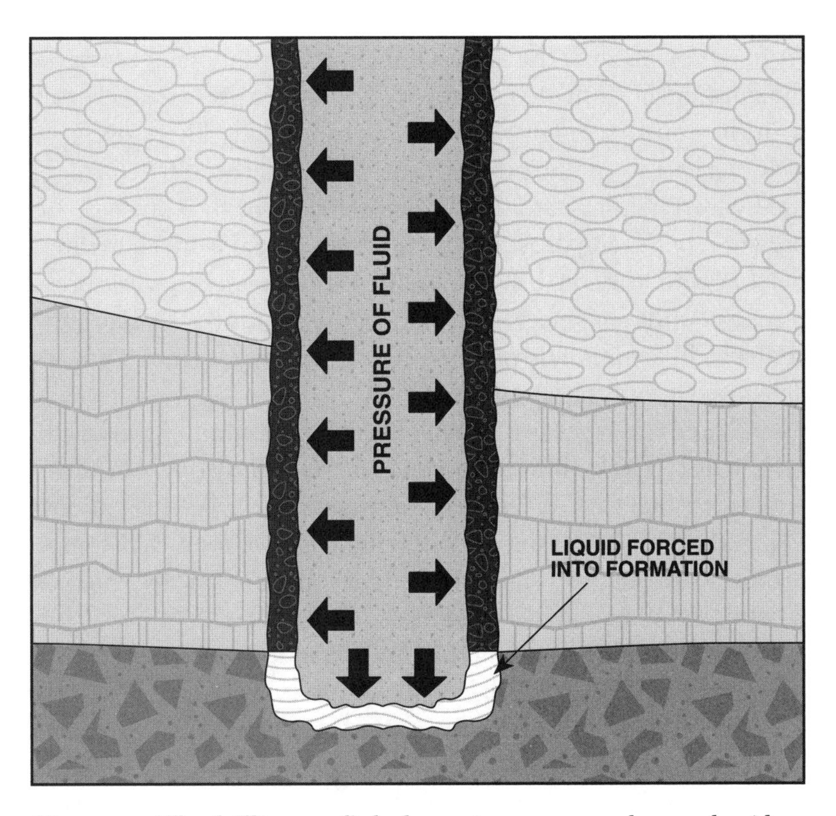

Figure 2. The drilling mud's hydrostatic pressure pushes on the sides and bottom of the hole, and forces the liquid part of the drilling mud into the formation.

Figure 3. Solid particles in the drilling mud plaster the wall of the hole and form an impermeable wall cake.

In most cases, wall cake is only a few thirty-seconds of an inch (mm) thick. Even so, a good filter cake not only stabilizes the wall of the hole, but also it slows the loss of liquid, called *fluid loss*, from the mud to a very low rate. If the fluid loss is not reduced to a low rate, crew members would have to constantly add more liquid to the mud to replace that lost to the formations.

A good filter cake is also thin and slick. It should be thin and slick so that if the drill stem contacts it, as it might in a crooked section of hole, the drill stem will not tend to get stuck in the wall cake. Crew members may add finely ground clays, such as bentonite, or other substances to drilling mud to improve its ability to form a filter cake, thus improving its wall-building ability.

Preventing Entry of Formation Fluids into the Well

Hydrostatic pressure is the force a fluid that is not moving exerts on the sides and bottom of a container. Hydrostatic pressure increases with the density (weight) and depth of the fluid. Therefore, in a well, the density and depth of the drilling fluid column determine the hydrostatic pressure. Both liquids and gases have hydrostatic pressure, but the hydrostatic pressure of a liquid is much greater. Persons who work on drilling rigs usually express hydrostatic pressure in pounds per square inch (psi) or kilopascals (kPa). The crew controls the hydrostatic pressure of drilling mud by increasing its density, or weighting it up.

All fluids exert pressure, including any gas, oil, and water inside a formation. The hydrostatic pressure of the drilling fluid can be the same as, greater than, or less than the pressure in drilled formations. When the pressure of the drilling fluid in the wellbore is the same as the pressure in the formation, drilling personnel say that the hole is *balanced*. When drilling fluid pressure is greater than formation pressure, the hole is *overbalanced*. And when formation pressure is greater than wellbore pressure, the hole is *underbalanced*.

In many cases, an underbalanced hole can be a problem because it allows formation fluids to flow into the hole. The uncontrolled flow of formation fluids into the wellbore is a *kick*. If crew members fail to detect a kick and do not take proper steps to control it, the well may blow out. A *blowout* is the uncontrolled flow of formation fluids to the surface (or into another underground formation). Blowouts are undesirable because they endanger the rig and its crew, waste precious resources, and can cause pollution. Because air and gas do not exert much hydrostatic pressure, they are not a good choice in a well where the operator expects formation pressures to be high.

On the other hand, an overbalanced hole can also cause problems. For one thing, a drilling mud whose weight is too high can create so much pressure on bottom that it fractures the formation. Depending on how badly the formation fractures, all or part of the drilling fluid can be lost to the formation. This *loss of circulation* means that none or only part of the drilling fluid returns to the surface.

Another problem with too much overbalance is that it holds cuttings on bottom. Although the jets of mud blast away at the cuttings, the too-high pressure keeps the jets from doing their job. Consequently, the penetration rate drops because the bit redrills old cuttings that hydrostatic pressure holds on bottom.

Ideally, the operator uses a mud whose weight develops just enough pressure to keep formation fluids from intruding into the hole. Although the mud weight may not balance formation pressure exactly, it overbalances it only slightly. Overbalancing only a small amount minimizes the possibility of lost circulation and maximizes the bit's ROP.

Powering Downhole Equipment

Some drilling operations use equipment in the hole that is not powered by a top drive or rotary table. For example, directional drilling uses a downhole motor to turn the bit because turning the whole drill stem from the surface is impossible (fig. 4). Shaped like a piece of pipe, a downhole motor can have turbine blades, or it can have a spiral steel shaft that turns inside an opening in the housing.

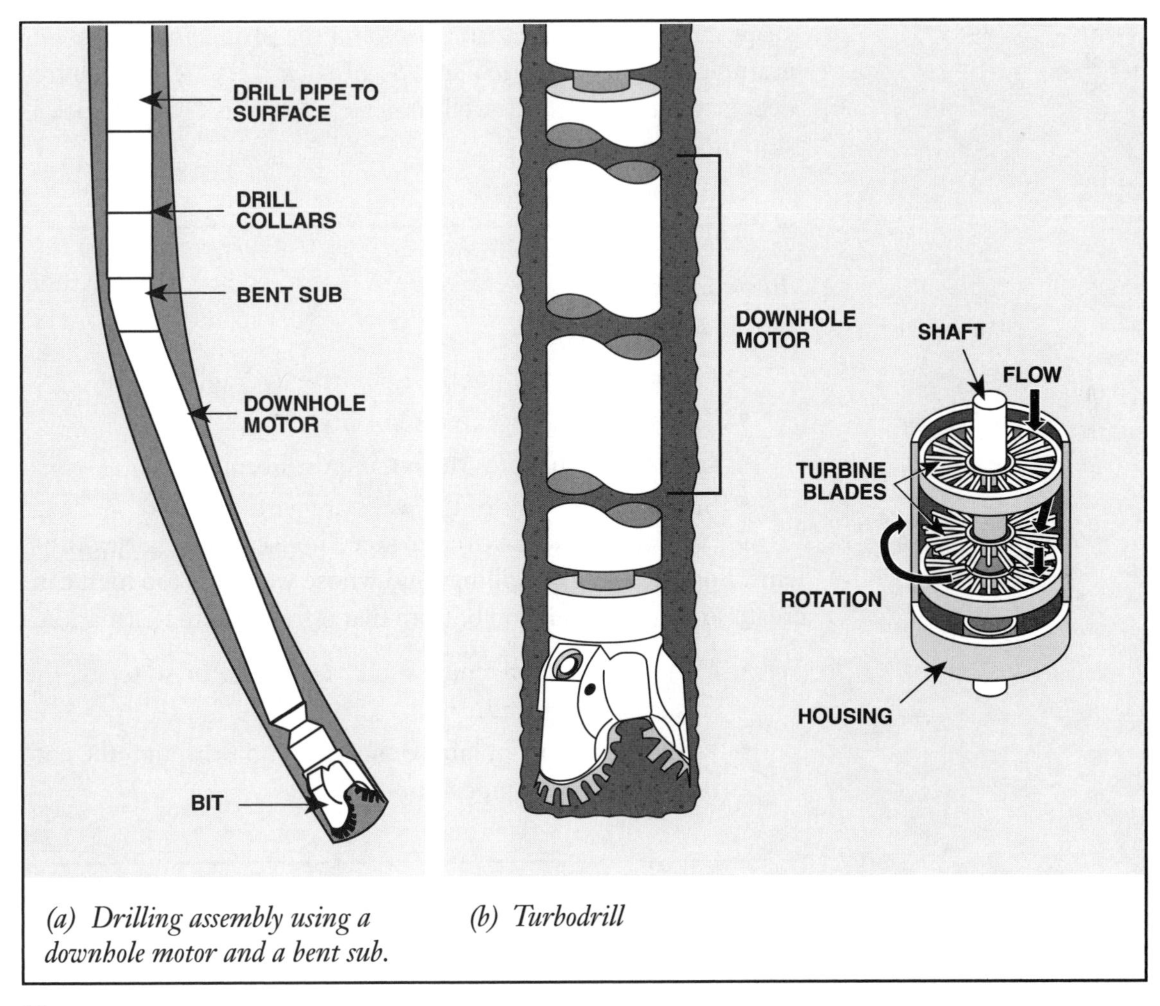

(a) Drilling assembly using a downhole motor and a bent sub.

(b) Turbodrill

Figure 4.

In the turbine type, the drilling fluid circulating inside the tool moves the turbine blades, which turns a shaft. The bit is attached to the shaft, so it turns when the shaft turns. With the spiral type, drilling mud flows through the opening in the housing and forces the spiral shaft to rotate. Again, the rotating shaft turns the bit.

Getting Information about the Formation Rock and Fluids

A geologist at a drill site periodically examines the cuttings to determine what formations are being drilled. Mud engineers also test drilling mud to see how much water, oil, or gas is entering the wellbore. If the well is a wildcat (the first well drilled in an area), they need as much information as possible to decide whether to continue drilling. When drilling in a known area, they still test the cuttings and fluid and compare the results with records from previous wells. Changes can reveal information about the shape and extent of the formations and the producing zones. For example, finding a particular type of rock a few feet or metres deeper than in a nearby well indicates that this layer of rock is sloping downward.

To summarize—

Functions of the circulation system

- Cleans the bottom of the hole with hydraulic power
- Transports cuttings to the surface
- Cools and lubricates the bit and drill stem
- Supports the walls of the wellbore with wall cake
- Prevents formation fluids from entering the wellbore by exerting hydrostatic pressure
- Transmits hydraulic power to downhole equipment
- Carries formation fluids such as oil, gas, or water to the surface for identification
- Reveals information about the formation through the cuttings and mud composition

---

▼
▼
▼

Drilling Fluid Composition

▼
▼
▼

The type of drilling fluid an operator selects to drill a well depends on many factors, including the type of formation being drilled, whether the formations contain water, the pressure the formations exert in the hole, and how deep the hole is. Because operators usually want to drill as fast as possible, they use the lightest fluid they can, but one that nevertheless controls downhole conditions. Air and gas provide the highest ROP. Liquid based muds (those that are mainly water, oil, or a combination of oil and water) usually cannot provide an ROP as high as air or gas, for reasons discussed shortly. Because air or gas cannot successfully drill most formations, the most common drilling fluid is drilling mud.

The relationship between the functions, the composition, and the properties of a drilling mud is close. Indeed, they are so closely related that it is difficult to separate them. The mud engineer has to decide which functions are most important in a particular well, and then choose the correct mud composition that has the properties to carry out those functions. Designing the mud program is a balancing act between the relative importance of one function over another, one composition over another, and one property over another to get the best performance for each hole.

The Mixture

Drilling mud is a three-phase fluid (fig. 5). Three-phase means that it is composed of three types of material. Drilling mud consists of a liquid phase and two solid phases. The solid phases are *suspended* in the liquid phase, which means that the particles are dispersed evenly throughout the liquid.

The liquid phase, or *continuous phase*, may be either water or oil. The two solid phases are the *dispersed phase*, because they are scattered or dispersed throughout the liquid. One solid phase consists of microscopic particles that react with the liquid. This phase is the *reactive*, or *colloidal*, *phase*. The main reactive solids in most drilling muds are clays.

The other solid phase is made up of solids that do not react with the liquid, the *nonreactive phase*. The nonreactive solids in drilling mud include finely ground cuttings, or *drilled solids*, such as sand, chert, limestone, dolomite, some shales, and mixtures of many minerals. (Some drilled solids are reactive, however. Reactive drilled solids include formations that are composed of or contain clay.) Nonreactive drilled solids are undesirable because they are abrasive and can damage pump parts. The mud may also contain nonreactive solids added purposely to alter its properties. One example of a nonreactive solid purposely added to the mud is barite, which is a weighting material. Barite increases the density of the mud without reacting with the liquid phase.

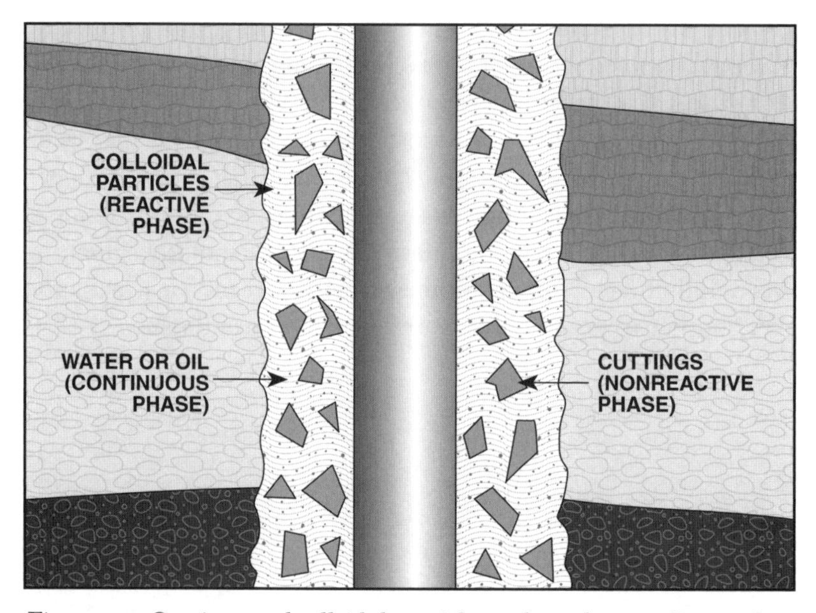

Figure 5. Cuttings and colloidal particles such as clay are dispersed throughout the water or oil that makes up a drilling mud. Colloids are actually too small to see without magnification.

Solids Content

Solids content refers to both the reactive and nonreactive solid phases of a mud—that is, the drilled solids and added solids. Mud characteristics such as density, viscosity, gel strength, yield point, and filtration depend on the type and amount of solids in the mud. Knowing the mud's solids content, because of its effect on these properties, can be helpful in diagnosing mud problems and planning treatments. For example, if the amount of solids in a high-viscosity mud is also high, adding water rather than chemicals may be the best way to thin it. The size and specific gravity of the solids are two characteristics that can affect the mud's properties.

Size

The American Petroleum Institute (API) classifies the solids in mud by their size. One measure of solid size is to pass a liquid containing solids of various sizes through a screen that has openings of a particular size. Some of the solids pass through the screen openings with the liquid and some do not. Whatever solids that remain on the screen are, of course, larger than the solids that passed through. Screen opening sizes are usually measured in terms of mesh—that is, the number of openings per square inch (in.^{2}) of screen material. Therefore a 50-mesh screen has 50 openings per in.^{2}, a 100-mesh screen has 100 openings per in.^{2}, and so on. At any rate, API considers as *sand* any particle that does not pass through a 200-mesh screen, regardless of what it is made of. Therefore, "sand" in this case may be particles of any sort of rock as well as actual sand. Sand is larger than 74 microns (μ) and smaller than 2 mm. (Micron is another term for micrometre, which is 1 millionth of a metre or about $^4/_{10,000}$ in.) Sand feels gritty between the fingers. Particles smaller than sand are called *silt*. Silt is so small it feels soft between the fingers. However, any drilled solids larger than 10 to 15 μ can be abrasive, so both sand and silt can damage equipment. The very smallest particles are *colloids*, less than 2 μ in size. They are invisible as individual particles. Because of their size, colloids have special properties—they do not settle out of the mud by gravity and they are small enough to pass through filter membranes.

Specific Gravity

Specific gravity is a number that indicates how dense a solid is compared to water. Water has a specific gravity of 1, so anything with a specific gravity of less than 1 is lighter than water and floats in it, and anything with a specific gravity of more than 1 sinks in it (fig. 6). Solids in mud have various specific gravities, most of which are higher than 1, but mud engineers generally classify them as either low-gravity or high-gravity, with low-gravity solids being lighter than high-gravity solids. Drilled solids and clays are usually low-gravity solids. On the other hand, crew members may add high-gravity solids to mud. One high-gravity solid, for example, is barite, a material for weighting mud. The specific gravity of the solids is an index to the relative amounts of clay and weighting material in the mud, information that is particularly valuable in controlling heavy muds.

Figure 6. The specific gravity of cork is less than that of water, so it floats, but a chunk of rock sinks. Even though most rocks are denser than water, they may be high- or low-gravity, depending on how much denser they are.

Additives

Anything a crew mixes into the basic circulating fluid is an *additive*. The additives in a drilling fluid to control its properties are many and varied. In a water-base mud containing additives, the proportion of additives changes as the mud circulates, so the derrickhand or a mud engineer tests the mud frequently to see whether to adjust the amount and composition of chemical additives. The composition of the mud may also need to be altered as the bit reaches new formations, each of which can present different problems.

To summarize—

Composition of mud

- Three phases
    - Continuous (liquid)
    - Reactive or colloidal (solid)
    - Nonreactive (solid)

Solids content

- The amount of reactive and nonreactive solids in a drilling mud
- Size ranges from smaller than 2 microns (colloids) to 74 microns (sand) and larger
- The specific gravity of the solids helps determine the relative amounts of clay and weighting material in the mud.

Basic Properties of Drilling Muds

▼
▼
▼

Some important properties of drilling mud are its density (weight), its viscosity, its gel strength, and its filtration properties. The effectiveness of the mud in performing its functions is directly related to these properties. Two problems in mud control are (1) determining what adjustments need to be made to the mud to give it the desired properties and (2) choosing how to make those adjustments.

Density (Weight)

In conventional rotary drilling, one of the chief functions of a drilling mud is to keep formation fluids, such as oil, gas, and water, in the formation. Under normal drilling conditions, preventing these fluids from flowing into the wellbore is crucial. An exception is when operators drill underbalanced. In this technique, also called drilling while producing, the operator deliberately keeps hydrostatic pressure below formation pressure, which allows formation fluids to flow into the wellbore while drilling. Operators primarily use underbalanced drilling in high-pressure but low-volume gas formations. A special sealing device at the surface—a rotating blowout preventer, or rotating head—vents the drilled gas through a special line away from the rig. Underbalanced drilling allows fast penetration rates because hydrostatic pressure is low and hole cleaning is therefore very efficient, which allows the bit cutters to constantly drill fresh, uncut formation.

When not deliberately drilling underbalanced and formation fluids unexpectedly flow into the wellbore, personnel say that the well has kicked. If crew members fail to control a kick, the kick can become a blowout. A blowout is the uncontrolled flow of formation fluids into the atmosphere or into another formation exposed to the wellbore. Uncontrolled flow of formation fluids into another formation is an underground blowout and can be very difficult to control. The density, or weight, of the mud, and its height, or length, determines the hydrostatic pressure of the mud column, and the correct hydrostatic pressure prevents kicks.

Mud that is Too Heavy

On the other hand, mud that is too heavy may fracture (break) the formation. Mud can flow into a fractured formation. Mud being lost to a formation is called lost circulation (fig. 7). Do not confuse lost circulation with fluid loss. As you recall, fluid loss is the relatively small amount of the liquid part of the mud (the filtrate) that hydrostatic pressure forces into a porous and permeable formation opposite the wellbore. Fluid loss leaves an impermeable, relatively thin sheath of solids on the wellbore, which is called wall cake. Lost circulation is the loss of actual drilling mud into a fracture. The loss may be partial, where some quantity of mud returns to the surface; or, it may be total, where none of the mud returns to the surface.

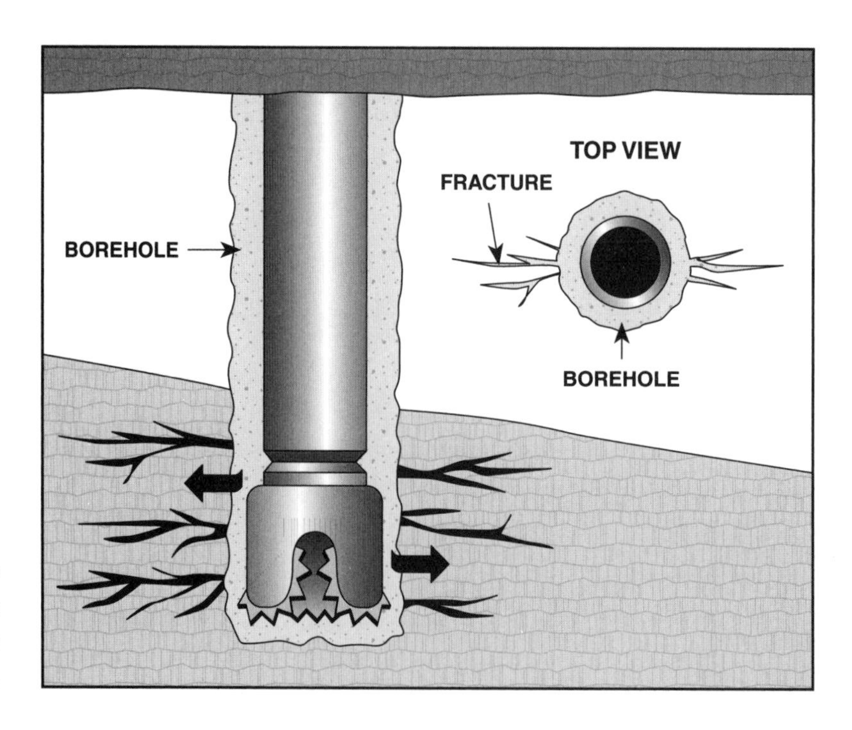

Figure 7. A mud that is too heavy can fracture the formation rock and start flowing into the cracks.

The balance between normal circulation and lost circulation can be very close. For example, a mud weight range of only 0.1 to 0.2 pounds per gallon (ppg) or 12 to 24 kilograms per cubic metre (kg/m^3) can mean the difference between adequately confining formation fluids and fracturing the formation. For example, a 14.5-ppg (1,740-kg/m^3) mud may control formation pressure, but a 14.6-ppg (1,752-kg/m^3) may break it down and cause partial or total lost circulation.

Even if excessive mud weight does not fracture a formation, it can slow ROP. Hydrostatic pressure that is too high tends to hold the cuttings on the bottom so that they cannot easily move up the hole—hole cleaning is inefficient. As a result, the bit redrills a lot of cuttings instead of fresh formation, which slows ROP.

Checking Mud Weight

Regardless of the weight needed, the derrickhand routinely checks the mud weight, along with other characteristics, to make sure that the mud weight is adequate and is doing its job. A sudden drop in mud weight, especially when accompanied by a gain in the level of mud in the tanks (pits) are two very obvious signs that the well may have kicked. Checking mud weight also ensures that it has not gotten too heavy.

Barite

Naturally occurring iron oxides such as hematite were among the first materials used to increase the weight, or density, of drilling fluids. In some cases, hematite is still used today. However, barite (sometimes spelled baryte), or barium sulfate ($BaSO_4$), is the most common weighting material in all types of treated mud because of its low cost, high specific gravity, inertness, and purity. Barite is a fine grayish-white powder with a specific gravity of 4.2 to 4.35, which means that it is 4.2 to 4.35 times denser than water. For example, if a given quantity of water weighs 10 lb or 4.5 kg, then the same quantity of barite weighs from 42 to 43.5 lb or 19.1 or 19.7 kg. Barite does not hydrate (take on water) when wet, which means that it does not react with water. It can be half to two-thirds of the cost of drilling mud when downhole pressure is abnormally high, but only about 5 percent of the cost in a normal well.

Table 1 shows the number of 100-lb sacks of barite that should be added to 100 barrels (bbl) of mud to achieve desired mud weights in ppg.

Table 1
Number of 100-lb Sacks of Barite per 100 bbl of Mud for Desired Mud Weight in ppg

Desired Weight (ppg)	100-lb sacks barite per ppg-increase per 100 bbl mud
9.5	57.2
10.0	58.3
10.5	59.5
11.0	60.7
11.5	61.9
12.0	63.2
12.5	64.6
13.0	65.9
13.5	67.4
14.0	69.0
14.5	70.6
15.0	72.3
15.5	74.1
16.0	76.0
16.5	77.9
17.0	80.0
17.5	82.2
18.0	84.5

Table 2 shows the number of 45.4-kg sacks of barite that should be added to 10 m^3 of mud to achieve the desired mud weights in kg/m^3.

Table 2
Number of 45.4-kg Sacks of Barite per 10 m^3 of Mud for Desired Mud Weight in kg/m^3

Desired Weight (kg/m^3)	45.4-kg sacks barite per kg/m^3-increase per 10 m^3 mud
1,140	35.75
1,200	36.4
1,260	37.2
1,320	37.9
1,380	38.7
1,440	39.5
1,500	40.4
1,560	41.2
1,620	42.1
1,680	43.1
1,740	44.1
1,800	45.4
1,860	46.3
1,920	47.5
1,980	48.7
2,040	50.0
2,100	51.4
2,160	52.8

Another weighting material is galena, which is a lead-bearing mineral with a specific gravity of 7.5. Keep in mind, too, that minerals such as calcium carbonate, sodium chloride, and other inorganic salts also add weight to muds, especially clear-water muds. In general, crew members can add inorganic salts to water and obtain mud weights up to about 10 ppg (1,200 kg/m^3).

Controlling Mud Weight

The common rocks and minerals found in formations—clays, shales, sand, and limestone—have specific gravities of about 2.6. The bit creates cuttings of these rocks, and, as the mud exits the well and flows over the shale shaker, the shaker removes most of them before they are recirculated back down the hole. Some of the cuttings are, however, so fine that they fall through the shaker screen. These fines enter the mud and, if not removed, add to the mud's weight, which may be undesirable. Operators and contractors may install desanders, desilters, mud cleaners, and centrifuges (fig. 8) to remove these solids from the mud. In some cases, where control of mud properties is not critical to successfully drilling a portion of the well, the crew can add water to the mud to adjust its weight to the required value.

Figure 8. Desander (a) and desilters (b) remove fine particles from drilling mud.

Table 3 shows how many bbl of water to add to 100 bbl of mud to reduce mud weight to a desired amount. Table 4 shows how many m³ of water to add to 10 m³ of mud to reduce mud weight to a desired amount. In many cases, though, adding water affects other mud properties, such as viscosity and gel strength, which can be detrimental.

Table 3
Effect of Water on Mud Weight (English Units)

Water Added (bbl/100 bbl mud)	Weight of Resulting Mud (ppg)							
0	10.0	11.0	12.0	13.0	14.0	15.0	16.0	17.0
5	9.9	10.9	11.8	12.8	13.7	14.7	15.6	16.6
10	9.8	10.8	11.7	12.6	13.5	14.4	15.3	16.2
15	9.8	10.6	11.5	12.4	13.3	14.1	15.0	15.9
20	9.7	10.6	11.4	12.2	13.1	13.7	14.7	15.6
25	9.7	10.5	11.3	12.1	12.9	13.7	14.5	15.3
30	9.6	10.4	11.1	11.9	12.7	13.5	14.2	15.0
35	9.6	10.3	11.0	11.8	12.5	13.3	14.0	14.7
40	9.5	10.2	10.9	11.7	12.4	13.1	13.8	14.5
45	9.5	10.2	10.9	11.6	12.2	12.9	13.6	14.3
50	9.4	10.1	10.8	11.4	12.1	12.8	13.4	14.1
60	9.4	10.0	10.6	11.2	11.9	12.5	13.1	13.7
70	9.3	9.8	10.4	10.9	11.5	12.0	12.6	13.1
80	9.3	9.8	10.4	10.9	11.5	12.0	12.6	13.1
90	9.2	9.7	10.3	10.8	11.3	11.8	12.4	12.9
100	9.2	9.7	10.2	10.7	11.2	11.7	12.2	12.7

Table 4
Effect of Water on Mud Weight (SI Units)

Water Added (m^3/15.9 m^3 mud)	Weight of Resulting Mud (kg/m^3)							
0	1,200	1,320	1,440	1,560	1,680	1,800	1,920	2,040
0.80	1,188	1,308	1,416	1,536	1,644	1,764	1,872	1,992
1.59	1,176	1,296	1,404	1,512	1,620	1,728	1,836	1,944
2.39	1,176	1,272	1,380	1,488	1,596	1,692	1,800	1,908
3.18	1,164	1,272	1,368	1,464	1,572	1,644	1,764	1,872
3.98	1,164	1,260	1,356	1,452	1,548	1,644	1,740	1,836
4.77	1,152	1,248	1,332	1,428	1,524	1,620	1,704	1,800
5.57	1,152	1,236	1,320	1,416	1,500	1,596	1,680	1,764
6.36	1,140	1,224	1,308	1,404	1,488	1,572	1,656	1,740
7.16	1,140	1,224	1,308	1,392	1,464	1,548	1,632	1,716
7.95	1,128	1,212	1,296	1,368	1,452	1,536	1,608	1,692
9.54	1,128	1,200	1,272	1,344	1,428	1,500	1,572	1,644
11.13	1,116	1,176	1,248	1,308	1,380	1,440	1,512	1,572
12.72	1,116	1,176	1,248	1,308	1,380	1,440	1,512	1,572
14.31	1,104	1,164	1,236	1,292	1,356	1,416	1,488	1,548
15.90	1,104	1,164	1,224	1,284	1,344	1,404	1,464	1,524

Gas cutting, which occurs when small amounts of natural gas from a drilled formation enter the mud, can lower the mud weight undesirably. As long as the gas is removed from the mud before it is recirculated back down the hole, gas cutting is usually not a problem because the amount of gas entering the mud is small. Bear in mind that the gas enters the well downhole, where the hydrostatic pressure is at its greatest. Then, when the gas and mud reach the surface, where the pressure is low, the gas expands and typically makes the mud look frothy or bubbly. To keep the mud weight at its correct value, however, and to prevent the mud pumps from gas locking, the gas must be removed before the mud is recirculated. (Gas locking occurs when the gas breaks out of the mud as it goes through an intake or discharge valve on the mud pump. The gas holds the valves partially open and pump efficiency is greatly reduced.) Usually, crew members install a device called a degasser (fig. 9) in the mud system to remove the small volumes of gas associated with gas cutting.

Figure 9. A degasser removes entrained gas from drilling mud.

Viscosity, Gel Strength, and Yield Point

Viscosity, gel strength, and yield point all have to do with how the mud flows and together they make up the mud's *flow properties*. The flow properties of the mud considerably influence the hydraulic output (the velocity and pressure of the moving fluid) at the bit, as well as the transport of cuttings up the annulus.

The *viscosity* of a drilling mud is its resistance to flow. A mud with high viscosity resists flow more than a mud with low viscosity, much as honey resists flow more than water. A viscous mud can transport more and heavier cuttings, so mud often contains a material to increase its viscosity. Viscosity must, however, be controlled, for a mud too viscous puts undue strain on the mud pump and may interfere with other desirable mud properties required to efficiently drill the well.

Drilling mud also has additives that cause the mud to gel, or stiffen, when circulation stops. This stiffening locks the cuttings in place throughout the mud column instead of allowing them to fall to the bottom of the hole. When the driller starts the mud pump, pump pressure reliquefies (ungels) the mud and it flows normally. *Gel strength* is a measure of the mud's ability to suspend the cuttings. Operators control gel strength with additives such as bentonite, which is a special clay that not only builds gel strength, but also viscosity. Gel strength must be controlled because a mud with a gel strength too high may require high pump pressure to restore flow (break circulation) after the mud has been in the hole only a short time.

The mud's *yield point* is a measure of the force required to start it to flow. The attraction between clay particles in the mud affect the amount of force required to get it to flow. This attraction causes the particles to *flocculate*, or clump together. This can be desirable or not, and it may be the result of either additives or contaminants. The operator might add a thinner to deflocculate a contaminated mud, or a flocculating agent to increase flocculation and thicken it.

For many reasons, viscosity, gel strength, and yield point should be carefully controlled. If they are too high, cuttings, sand, and entrained gas cannot easily escape or settle out. Moreover, high-viscosity muds require high pump pressures, which increase wear and tear on the pump. What's more, the more viscous a mud is, the easier it is for the crew to swab in formation fluids when pulling pipe out of the hole. Similarly, the more viscous a mud is, the easier it is for the crew to create pressure surges when running the pipe into the hole. These surges, if high enough, can fracture a formation and create lost circulation.

Increasing Viscosity with Clays

Operators add clay to many drilling muds—the common oilfield term for adding clay is to *mud up*. The important characteristic of clays, for drilling mud purposes, is that they absorb water and *hydrate*, or swell, and form a gel (fig. 10). The swelling and gelling qualities of clays increase the viscosity of the mud and build filter cake to reduce filtration.

Bentonite, a commonly used colloidal clay , is mostly made up of montmorillonite, a clay mineral that disperses into small particles in water and hydrates. Natural clays (those found when drilling) sometimes make a good mud, sometimes hydrate only slightly, and sometimes hydrate a great deal and make the mud very viscous. When the makeup water is salty, bentonite does not work as well, so the mud engineer may use a saltwater clay, such as attapulgite.

Other additives that increase viscosity include asbestos and various polymers. Asbestos is very effective in both freshwater and saltwater muds, but is a carcinogen, so its use is controlled. *Polymers* are a class of chemicals that may be synthetic or natural. Polymer additives to mud include *carboxymethyl cellulose (CMC)* and starches. They form even more powerful colloids and gels than bentonite but they dissolve in water instead of being suspended in it as fine particles. So polymers increase the mud's viscosity without added solids, and the mud engineer may decide to add less bentonite along with a polymer. Polymers often affect other properties of the mud than viscosity, so they serve more than one function.

Figure 10. Bentonite reacting with water

Clay Yield

Clay yield refers to the number of bbl (m^3) of mud with a viscosity of 15 centipoises (cp) that 1 ton (tonne) of clay can produce. The defining value for yield is 15 cp because the critical part of the clay yield curve for all types of mud appears at that point (fig. 11). Additions of clay up to 15 cp promote little viscosity. Note for example in figure 11 the curve for a low-yield clay. It takes a little less than 150 pounds per bbl (lb/bbl), or a little less than 434 kg/m^3, to achieve a mud with a viscosity of 10 cp. Similarly, to achieve a mud viscosity of 15 cp, it takes about 200 lb/bbl (579 kg/m^3). At this point, however, the clay yield curve rises rapidly, so that the addition of only 50 more lb/bbl (145 kg/m^3) of clay increases the viscosity to 60 cp, a four-fold increase.

Knowing the clay yield gives other useful information about a mud. For example, according to figure 11, about 20 lb/bbl (58 kg/m^3) of Wyoming bentonite is required to produce a 15-cp mud. Further, such a mud contains about 6 percent solids by weight, yields 90 bbl/ton (15.75 m^3/tonne), has about 2.5 percent solids by volume, and weighs about 8.6 ppg (1,032 kg/m^3).

Decreasing Viscosity

Salt, cement, anhydrite (a naturally occurring mineral containing calcium sulfate found in some formations), and drilled solids can cause a mud to be too viscous. Also, a type of sticky, natural clay formation known as gumbo can react with the mud and thicken it. Flocculated clay solids can also cause high viscosity.

If drilled solids cause the mud's viscosity to be too high, using a fine-mesh shale shaker screen, desanders, desilters, and mud cleaners to mechanically separate them may be all that is necessary. Some types of mud contamination, however, such as that caused by salt, cement, and anhydrite, require thinning with water or chemicals. The disadvantage of adding water is that it lowers viscosity and reduces density. Thus, the crew must add more weighting material to raise the density. Chemicals do not reduce density, so when using weighted muds, the crew may add water, chemicals, or both. When the percentage of solids is in the correct range, they usually add chemicals. If the percentage of solids is high, they usually add water. To *water-back* a mud refers to diluting a mud with water.

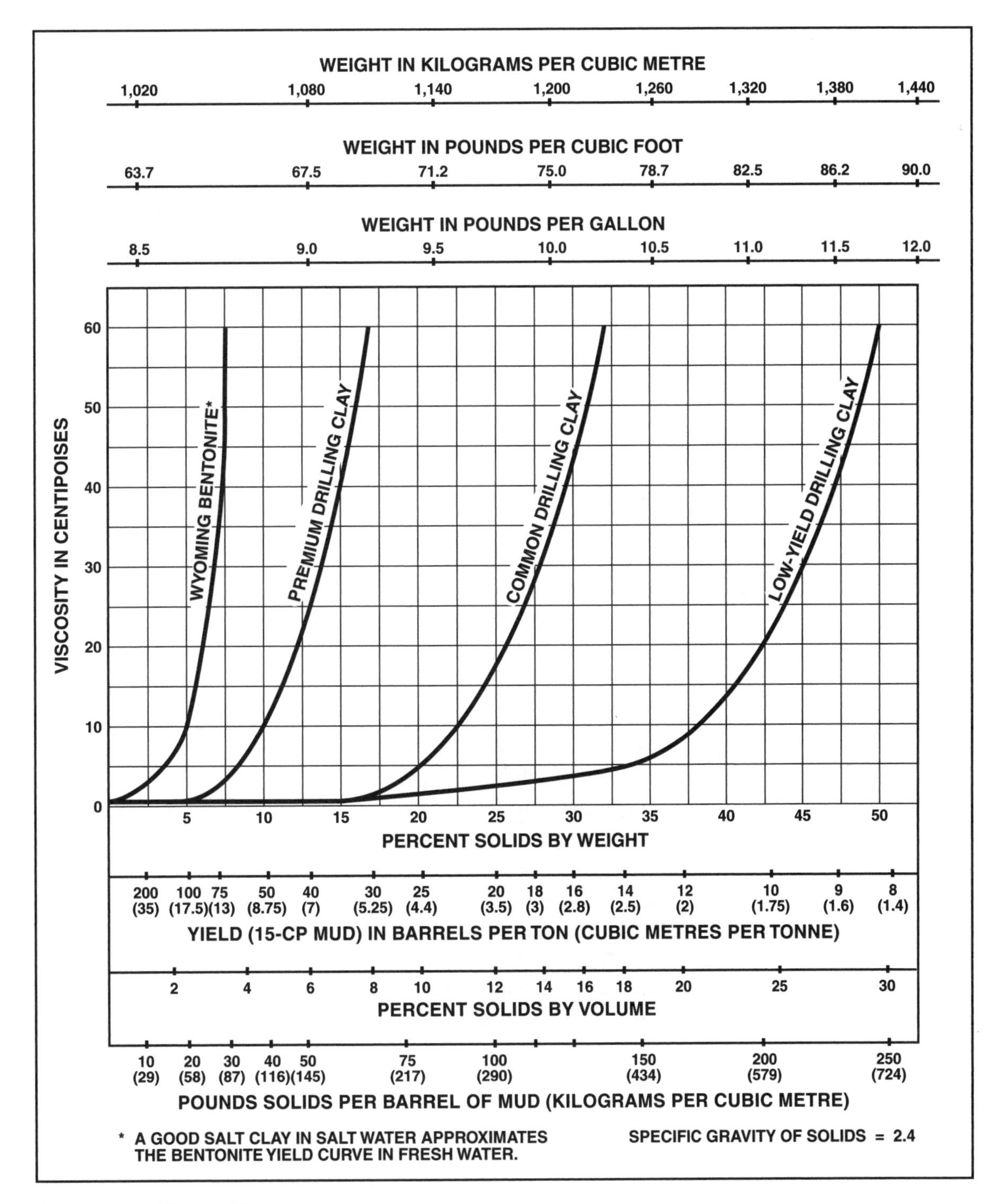

Figure 11. Clay yield curve

Viscosity-Reducing Chemicals

Chemicals used to reduce viscosity and gel strength include a group of compounds called *dispersants*. Dispersants include plant tannins, lignites, polyphosphates, and lignosulfonates. Lignosulfonates, the most widely used dispersant, are very effective but are also expensive. They can be used in all water-base mud systems. Lignites are the next most popular after lignosulfonates; phosphates and tannates are now rare.

Some dispersants do several other jobs as well, which may be more important than reducing viscosity. Both lignites and lignosulfonates reduce fluid loss and the thickness of the filter cake, counteract the effects of salts, minimize the effect of water on the formation, emulsify oil in water, and stabilize the mud at high temperatures.

Filtration

Filtration occurs when a liquid that contains undissolved solids flows through a very fine strainer like filter paper and leaves the solids behind. If you use a coffee maker that uses a filter, then you know what filtration is. The coffee machine pours hot water through the coffee into a container in which the filter is placed. The water runs through the ground coffee and into the pot below. In this case, the coffee is the filtrate, the liquid part of the drink that is free of solid coffee grounds.

When a mud-filled borehole is opposite permeable formations, the mud's hydrostatic pressure continually forces the liquid part of the mud (water or oil) into the formations. A permeable formation, such as sandstone, works like the filter paper; it holds back the solids in the mud but allows the liquid to pass into the rock's pore spaces. These solids are deposited onto the face of the sandstone in the form of a filter cake.

Keep in mind that the mud in the borehole has to be opposite a permeable formation for filtration to occur. Where mud is opposite a virtually impermeable formation, such as shale, no cake forms. In this case, the water in the mud wets the surface of the shale, where there is just enough permeability to let the water contact shale particles. This water contact can cause the shale to swell and slough into the hole. Whether the shale sloughs depends on how salty the water in the mud is compared to how salty the water in the pores of the shale is. If the salt content of the water is considerably less than that contained in the shale, then the shale may hydrate (take on water), swell, and slough.

Controlling Fluid Loss

The liquid part of the mud that flows into the formation is fluid loss—the water or oil lost from the mud. Excessive fluid loss has several undesirable effects. One is that as long as the fluid filters into the formation, the filter cake keeps getting thicker. It may eventually become thick enough to reduce the diameter of the hole, causing tight spots where the drill string can get stuck (fig. 12). Second, muds with a high fluid loss may sometimes cause sloughing and caving of shale formations because, as previously mentioned, some shales are sensitive to water. Third, in porous formations, high fluid loss can also hinder the interpretation of electric logs. The problem is that one formation characteristic that an electric log measures is the resistivity of the naturally occurring water in the pores of a rock. Filtrate, of course, is not naturally occurring; instead, it is water from the mud. Consequently, the log may measure the resistivity of the mud's water rather than the formation's water.

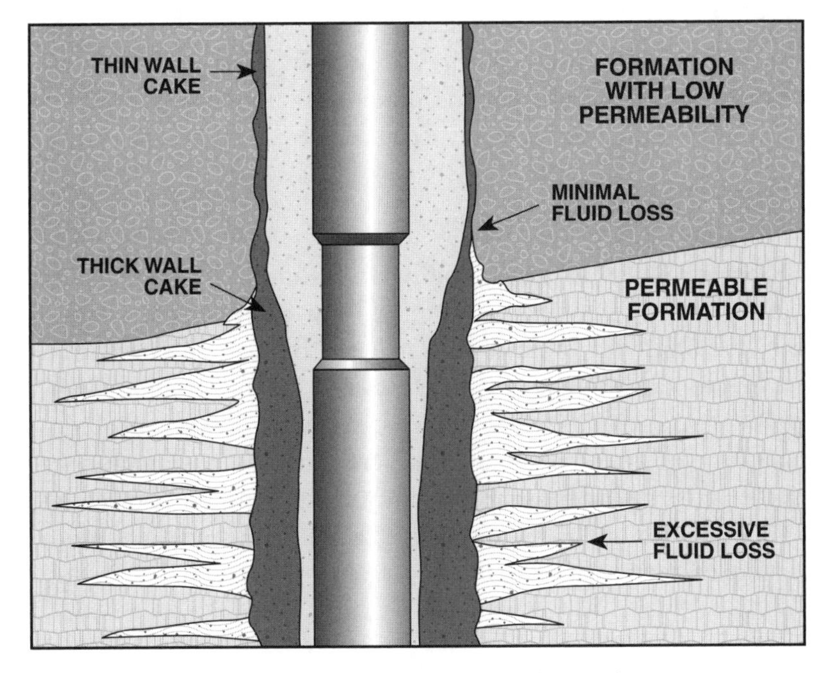

Figure 12. Fluid loss affects thickness of the filter cake.

Finally, liquid entering the producing zones (the zones containing hydrocarbons) may reduce the rate of oil flow when the well is ready to produce. This happens because, like shales, some producing zones are sensitive to water. The water enters tiny openings of the zone and causes the surrounding rock to swell, thus blocking permeability. This phenomenon is *formation damage*.

Additives that control fluid loss pack tightly together to form a good filter cake. Many of the materials that control fluid loss are used for controlling other properties, such as bentonite, lignites and lignosulfonates, CMC, synthetic polymers, and starches.

Inhibiting Swelling of Water-Sensitive Formations

When shales and clays swell and slough into the hole, they can impede drilling, enlarge the hole, increase viscosity, and damage producing zones. Adding lime, which is calcium hydroxide, or $Ca(OH)_2$; gypsum, which is calcium sulfate, or $CaSO_4$; calcium chloride, which is $CaCl_2$; or some other calcium compound to inhibit this swelling was common until the 1970s. Calcium muds, also called *lime muds* or *gyp muds*, are still in use, but today most such drilling operations use an oil mud or a nondispersed water-base mud containing a salt or special additives.

pH Value

The pH of a fluid is its relative acidity or alkalinity. Double-distilled water is exactly neutral—neither acid nor alkaline. On the pH scale, the number 7 indicates this neutral point. Acids range from just below 7 for slight acidity to less than 1 for the strongest acidity. Alkaline solutions range from just above 7 for slight alkalinity to 14 for the highest alkalinity. In increasing order, each number on the pH scale indicates an alkalinity of ten times the preceding number. For example, a solution with a pH of 9 is ten times more alkaline than one with a pH of 8.

Desirable pH Values

Usually, mud must be alkaline, with a pH of between 8 and 13, to allow the chemicals in the mud to work well and to minimize corrosion. For example, some mud additives need an alkaline environment to work properly, and the detection and treatment of certain contaminants depends on pH. Acids accelerate corrosion, but a pH of 10 to 12 minimizes the corrosion rate.

Table 5 lists some common types of drilling muds and their usual pH range. Organic-dispersant treated muds—mud using quebracho, lignite, and lignosulfanates—are grouped together. The choice of dispersant is determined to some extent by the pH range desired and the type of mud used. Polyphosphates are ineffective at high temperatures and are rarely used in a mud with a pH higher than 10. Plant tannins such as quebracho are effective in muds with a pH above 8. The optimum pH range for lignites is between 8.5 and 9.5. They work well in low-pH emulsion muds and they are often used in lime muds. Calcium lignosulfonate was first used only in lime muds, but modified lignosulfonates are now used in almost all types of muds.

Table 5
The pH of Common Drilling Muds

Mud Type	pH
Low solids, salty	6.5 plus
Phosphate	7.5–9.5
Organic-dispersant treated	7.5–12.0
Gypsum	8.0–10.0
Calcium chloride	10.0 plus
Lime treated	11.5 plus

pH Values of Common Additives

Additives themselves can change the pH of a mud. Although the pH of one common additive, barite, is approximately 7.0, mud containing bentonite or most clays (viscosity increasers) usually has a pH of 8.5 to 9.5. Muds containing lime have a pH around 12. Table 6 lists the chemical names of the common mud additives and the approximate pH of a 10 percent solution of each chemical.

Table 6
The pH of Common Mud-Treating Agents

Treating Agent	pH
Barium carbonate, $BaCO_3$	10.3
Sodium bicarbonate, $NaHCO_3$, baking soda	8.3
Calcium sulfate, $CaSO_4$, ½ H_20, gypsum-plaster	6.0
Chrome lignosulfonates	3.4–4.0
Sodium carbonate, Na_2CO_3, soda ash	11.0
Calcium hydroxide, $Ca(OH)_2$, slaked lime	12.0
Sodium hydroxide, NaOH, caustic soda	13.0
Calcium lignosulfonate	7.0
Lignite	5.0
Quebracho	3.8
Sodium acid pyrophosphate, $Na_2H_2P_2O_7$	4.8
Sodium hexametaphosphate, $(NaPO_3)_6$	6.0
Sodium tetraphosphate, $Na_6P_4O_{13}$	7.5
Tetrasodium pyrophosphate, $Na_4P_2O_7$	9.9

Other Additives

The crew may add a number of other substances to improve the properties of the drilling mud. *Flocculants* cause drilled solids to stick together and clump into larger pieces that are easier to remove than very small particles. *Corrosion-control agents* reduce corrosion to the metal parts that salts in the mud can cause. *Defoamers* break up foams that form when formation gas contaminates the mud. *Lubricants* reduce friction, which reduces bit wear and allows the drill string to turn more easily against the wall of the hole. *Antidifferential sticking additives* help free stuck pipe. *Emulsifiers* improve lubricating properties, lower fluid loss, and help prevent the bit from balling up. Emulsifying water in oil or oil in water allows a lower pump pressure and a faster drilling rate in shales.

To summarize—

Basic mud properties include

- Weight or density
- Viscosity
- Filtration rate
- pH value

To control mud properties

- Barite is the usual additive to weight up drilling mud.
- Equipment that separates drilled solids or entrained gas from the mud helps control its weight.
- Clay is the usual additive to mud up, or increase viscosity, gel strength, and yield point of the mud.
- Clay yield is the number of bbl (m^3) of mud with a viscosity of 15 centipoises (cp) that 1 ton (tonne) of clay can produce.
- Mechanical separation of drilled solids and addition of water or dispersants can reduce a mud's viscosity.
- Additives that control fluid loss pack tightly together to form a good filter cake. In shales, using an oil mud or a non-dispersed water-base mud containing a salt or encapsulating agent controls fluid loss.
- Usually, drilling mud is alkaline (pH 8–13) to allow the chemicals in the mud to work well and to minimize corrosion.

▼
▼
▼

Composition of Water-Base Drilling Muds

▼
▼
▼

Water-base muds are the most widely used drilling fluid. Because either fresh water or salt water is available in most of the world's drilling areas, water-base muds are easier to use and less expensive than oil-base muds or compressed air.

The quality of water used to make up and maintain water-base muds affects the way mud additives perform. For example, clays work best in distilled water. Hard water—water that contains large amounts of calcium and magnesium salts—and salt water reduce a clay's effect.

Spud Muds

The composition of the mud used to start a well, or *spud in*, varies with drilling practices around the world. Sometimes the operator uses water alone from a nearby source such as a well, stream, or lake. The ideal situation is when the makeup water is soft and the formations near the surface make a good natural mud. If this is not the case, the operator may mix clay, lime, and soda ash into the mud. Lime thickens the mud and allows the operator to use less clay to build viscosity.

Natural Muds

Sometimes the formations near the surface contain enough clay to make up a good *natural mud* when mixed with water. Natural muds that have low weight and viscosity, because they either do not hydrate well or need a lot of water to keep the weight and viscosity down, are useful at shallow depths, such as for surface drilling and for making hole below the conductor casing. In shallow holes, the formation pressures are usually normal, and mud does not need to be heavy to prevent kicks. These low-weight, low-viscosity muds provide a high rate of penetration and decrease the risk of stuck pipe and lost circulation.

Operators have drilled wells to 12,000 ft (over 3,500 m) on the U.S. Gulf Coast using natural mud with a little commercial bentonite added. Usually, however, as the hole gets deeper, the crew adds other chemicals to natural mud to improve its properties.

Disposable Spud Mud

Some wells are spudded in through conductor casing already cemented in the cellar. When drilling through the conductor casing, the cement drilled out of the conductor contaminates the spud mud. So, after drilling past the conductor casing, the operator usually properly disposes of the spud mud because it is not suitable for deeper formations. In cases where the rig drills the conductor hole, the driller uses a spud mud and drills the conductor hole to the required depth. Crew members then run and cement the conductor casing. In such cases, the mud may not be discarded because it was not contaminated by cement.

Good-Quality Spud Mud

In some places, surface formations may consist of loose sand and gravel. In such cases, spudding in requires a fairly good quality of mud. This spud mud must be able to build up a good filter cake on the wall of the hole to prevent caving. It must also be viscous enough to carry cuttings out of the hole as it is drilled. Bentonite and attapulgite are common additives to spud muds to control viscosity. Sometimes the most economical option is to get spud mud from another drilling operation.

Mixing Spud Mud

Because mud usually moves up the annulus fairly slowly in large-diameter conductor holes, many operators prefer to make a thick mud that can more readily carry cuttings from surface sands and gravels. Generally, they add only bentonite or a premium clay to water to build viscosity. In a makeup water that is not too hard or salty, 20 sacks of bentonite can make about 80 to 100 bbl (about 13 to 16 m^3) of mud. Twenty sacks of premium clay can make about 45 to 65 bbl (about 7 to 10.5 m^3) of mud. Twenty sacks of a cheap drilling clay make about 10 to 21 bbl (about 1.5 to 3.5 m^3).

One crew member usually mixes the spud mud while others finish rigging up. To mix in the clay, the crew member runs water into a mud tank (pit) and circulates it through the mixing hopper. (Note that "mud pits" is the traditional term for the steel, usually rectangular, open tanks that hold active drilling mud on the rig. The term is a holdover from the old days when earthen pits excavated near the rig held the active drilling mud. The proper term, however, is tank and this book uses it throughout.) The crew member should take from 2 to 15 min to empty each sack into the throat of the hopper. Clay that balls up and floats on the surface of the tank indicates that it has been added too quickly. In this case, the crew member should activate the tank agitators to stir the mud in the tank. The crew member continues adding clay until the mud reaches a funnel viscosity of 30 to 35 seconds. Funnel viscosity is a measure of how long it takes for a quart (qt) or litre (L) of mud to flow through a Marsh funnel. This viscosity test is covered in detail later. The funnel viscosity of 30 to 35 seconds is a little lower than the mud's final viscosity because viscosity increases as the mud ages and picks up solids during drilling.

Chemically Treated Muds

Muds that have chemicals added to them to alter their properties go by a number of names, depending on the additive. Chemical additives generally react with the continuous phase of a mud—that is, they are mixed with the water, oil, or water and oil that make up the liquid part of the mud and react with it. Certain additives, however, are nonreactive. For example, barite, lost-circulation materials, and drilled solids like sand and silt. Table 7 lists the typical composition of each phase of drilling mud. Table 8 shows some of the basic chemicals used to treat muds. Chemical treatment usually prevents or corrects severe hole problems. However, chemicals that may be desirable for some hole conditions may be undesirable contaminants under other conditions. For example, calcium in drilled solids can be a contaminant that prevents clay additives from building viscosity. At the same time, however, a mud company may supply a calcium compound for the operator to use as an additive to inhibit shale sloughing.

Table 7
Composition of Drilling Muds

Phase	Composition
Continuous phase	Fresh or salt water Oil Water and oil
Reactive (colloidal) phase	Clay (usually bentonite) Dispersant, flocculant, corrosion control agent, pH control agent, defoamer (water-base muds) Emulsifier (oil muds)
Nonreactive phase	Drilled solids (sand and silt) Weighting material (usually barite) Lost circulation material

Table 8
Basic Mud Chemicals

Viscosifiers	*Swelling inhibitors*
Bentonite	Salt
Attapulgite	Encapsulating agent
Polymers	Lime
	Gypsum
Viscosity-reducing chemicals	*Emulsifiers*
Lignosulfonate	Lignites
Lignites	Lignosulfonate
Tannates	Detergents
Phosphates	
Fluid-loss reducers	*Lost-circulation materials*
Starches	Granular
CMC	Fibrous
Synthetic polymers	Flaked
Lignites	Slurries
Lignosulfonate	
Weighting materials	*Special Additives*
Barite	Flocculants
Hematite	Corrosion controller
Galena	Defoamer
Calcium carbonate	pH controller
Dissolved salts	Mud lubricant
	Antidifferential sticking material

Phosphate-Treated Muds

For shallow wells of 7,500 feet (2,000 metres) or less, phosphate treatment of a natural mud may be the primary or the only treatment. Phosphate treatment is usually limited to drilling areas that have the proper base materials for natural mud. Complex phosphates are very effective in reducing viscosity, gel strength, and filtration rate, even when used in such small quantities as 0.1 to 0.2 lb/bbl (0.3 to 0.6 kg/m^3) of mud. The crew may add small amounts of bentonite to improve wall-building and reduce fluid loss, but too much bentonite usually increases the viscosity in phosphate muds above a desirable amount.

Usually, operators do not use phosphate-treated muds for wells deeper than about 7,500 feet (2,000 metres), since complex phosphates cannot withstand temperatures above 200°F (93°C). Phosphate treatment is also ineffective in muds that are contaminated with salt or calcium.

Lignosulfonate Muds

To keep the viscosity, gel strength, and filtration low, some muds are treated with lignosulfonate, a chemical additive that is a by-product of papermaking. Lignosulfonates affect flow properties by deflocculating solids and clay particles. They coat the particles and keep them apart (fig. 13).

Some mud systems ordinarily require very little chemical deflocculation. However, when a contaminant such as anhydrite, cement, or salt water enters the system, it neutralizes the natural forces keeping the solid particles apart, and they flocculate. In such cases, lignosulfonates may be used as a deflocculant.

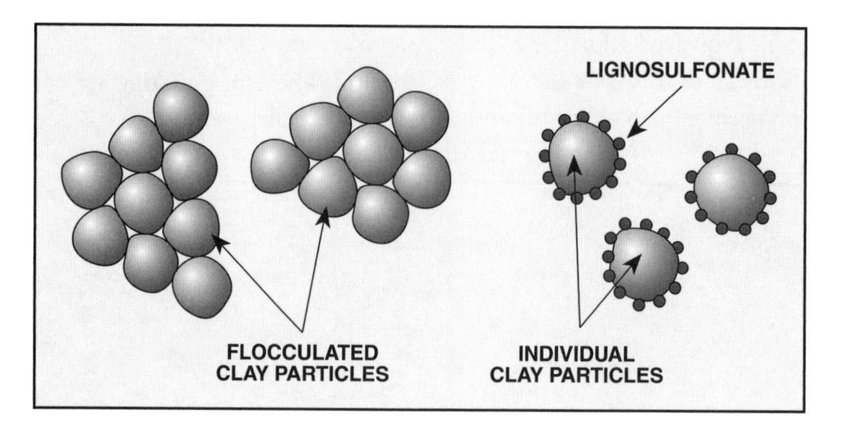

Figure 13. Lignosulfonate forms a protective coating around colloidal particles to keep them from clumping together.

Where Lignosulfonate Muds are Useful

Operators use lignosulfonate muds in deep wells that need heavy muds and have high bottomhole temperatures. Lignosulfonate keeps viscosity low even when the solids content is high from adding barite. And it is effective when bottomhole temperatures are as high as 400°F (204°C). Lignosulfonates work well in muds contaminated by calcium or salt. Lignosulfonates and soluble lignite compounds are also excellent emulsifiers when the crew must add oil to a water-base mud.

Amount of Lignosulfonate

The amount of lignosulfonate the crew adds to a mud varies widely. Depending partly on whether the water in the system is hard or soft, the amount can range from about 0.5 to 10 lb/bbl (1.5 to 29 kg/m^3). The mud engineer often begins with 1 lb/bbl (3 kg/m^3) or less in relatively shallow wells. Treatment with 3 to 6 lb/bbl (9 to 17 kg/m^3) further changes flow properties. For the lowest filtration rate, 10 lb/bbl (29 kg/m^3) of lignosulfonate may be necessary. In general, seawater muds require larger quantities than freshwater muds.

Calcium-Treated Muds (Lime Muds)

The addition of lime hydrate, gypsum, calcium chloride, or other calcium compounds to drilling mud water was very common in the late 1940s and 1950s. Lime muds are less common now, but, because of their low cost, they are still in use when hole conditions allow.

Muds that contain significant amounts of calcium and magnesium tend to suppress or inhibit the swelling of clays and shales that the mud contacts in the hole. This inhibiting property helps control shale sloughing, hole enlargement, excessive increases in viscosity due to dispersion of solids in the mud, and damage to clay-bearing productive sands.

The calcium used for treating a mud system may be obtained commercially as lime hydrate (slaked lime), gypsum, or calcium chloride. Calcium is also available from the drilling of soluble calcium beds such as anhydrite or from drilling through cement. The seawater used in offshore drilling fluids also contains calcium and magnesium in ample quantities for calcium treating.

Calcium-treated muds deflocculated with lignosulfonates have lower viscosities for a given solids content and more resistance to contamination by calcium and salts than muds treated with quebracho. And lignosulfonate muds are more stable and control fluid loss at temperatures over 300°F (149°C).

Disadvantages of Calcium-Treated Muds

Today, calcium-treated muds are not widely used. Lignosulfonate muds without added calcium have replaced them in deep, high-temperature wells because of a serious problem: severe gelation. This problem occurs over time at high temperatures in muds with high pH values, and is most severe in lime muds, especially red lime muds with a pH of 11.5 to 12.5. The combination of clays, calcium, and high pH reacting slowly at high temperature forms a sort of cement that causes the mud to set up or become so severely gelled that it cannot be circulated out. Gelation is minimized when lignosulfonate muds are used because the addition of calcium is unnecessary, and pH can be maintained in the 8.5 to 9.5 range to minimize the hydroxyl ion concentration.

Low-Solids Muds

Solid materials picked up in the course of drilling can cause serious problems. Unless they are removed from the drilling fluid, these drilled solids can damage mud pumps, bits, and other costly equipment. Drilled solids also reduce the hydraulic horsepower at the bit and increase mud weight, which lower ROP. Solids reduce hydraulic horsepower because the solids in the mud increase pressure losses within the mud itself. The solids rub together and increase friction, which leads to increased pressure losses. Increased mud weight can decrease ROP because an unnecessarily heavy mud creates more hydrostatic pressure on bottom that tends to hold down the cuttings on the bottom of the hole. The jets of drilling mud simply cannot move the cuttings out of the way as well as they can with lighter weight mud.

The term *low-solids mud* refers to any drilling fluid in which the amount of drilled solids is controlled. The purpose of a low-solids mud is to satisfy the demands of the well conditions while maintaining the highest possible drilling rate.

Air drilling excepted, drilling with clear, fresh water achieves the fastest ROP. Unfortunately, its use is limited because of its low weight, its tendency to soften shales and dissolve salts, and its poor filtration properties. Clear salt water, or brine, is more widely used. But both fresh water and salt water pick up solids during drilling. These drilled solids must be removed if clear water is the desired drilling fluid. Devices like shale shakers, desanders, mud cleaners, and desilters can remove the larger particles. Clays and other fine particles are more difficult to separate, often requiring chemical flocculants and a large settling tank for the aggregates to settle out.

Polymer Muds

A specific group of low-solids muds based on chemical compounds known as polymers has come into wide use. A polymer is a single large molecule that has a high molecular weight and is made up of repeating chemical units known as monomers. Polymer muds —muds that incorporate polymers—are effective for flocculating solids, increasing viscosity, controlling fluid loss, and stabilizing shale.

Polymer muds have another important property called shear thinning. *Shear thinning* is a reduction in viscosity that happens when the mud is under shear stress. *Shear stress* is the force applied to a liquid to make it flow. The amount of shear stress (the *shear rate*) varies in different parts of the circulation system. For example, the shear rate is high at the bottom of the hole and lower in the annulus. So a mud that reacts to high shear stress by thinning improves the drilling rate—since the viscosity is lower at the place where the bit is working—while still maintaining the ability to move cuttings up the annulus.

Bentonite Extenders

One group of polymers greatly increases the viscosity produced by bentonite, while flocculating other clay solids so that they are simple to remove from the mud. With these polymers, or *bentonite extenders*, the mud has the desired viscosity with only about half the bentonite that would normally be required. The shear thinning of these muds makes for efficient hole cleaning. The system remains flocculated, however, and, consequently, drilled solids do not pack tightly in the filter cake, causing filtration rates to be fairly high. Filtration-reducing additives can be used, but they may affect the shear-thinning characteristic.

Extended Bentonites

Another freshwater low-solids mud system uses *extended bentonites*, which are selected bentonites treated with chemical polymers. These extended bentonites yield about twice as much viscosity as regular bentonite, so the bentonite solids content is only half the usual amount. Adding a deflocculant to an extended bentonite mud can reduce the mud's viscosity and yield point. However, when compared to a mud with the same amount of untreated bentonite to which a deflocculant has been added, the viscosity and yield point of an extended bentonite mud is greater.

Extended bentonite mud has a greater shear-thinning property than a regular bentonite mud, making faster drilling possible. On the other hand, filtration rates with extended bentonites are nearly double those of regular bentonite muds.

The use of bentonite extender and extended bentonite muds is limited to areas where good fresh water is available. These muds are most effective when the calcium content of the water is low.

Biopolymers

When a particular strain of bacteria eats carbohydrates, it produces another widely used polymer, a *biopolymer*, or *X-C polymer*. A biopolymer is soluble in fresh or salt water in a pH range of 6 to 10. It produces large increases in apparent viscosity and yield point with moderately good filtration control.

Biopolymer muds can be treated to increase viscosity and shear thinning when the crew adds another chemical that crosslinks, or forms a bridge, with the biopolymer molecules. Chromic chloride or chrome alum are two of these crosslinking agents. Adding relatively small amounts of the biopolymer (0.5 to 1.0 lb/bbl, or 1.4 to 2.9 kg/m^3) and the crosslinking agent (about 20 percent of the biopolymer volume) to the mud system can produce high viscosity in the annulus and very low viscosity at the bit, which lifts the bit cuttings well while keeping fast drilling rates. Biopolymer muds also allow desanders and desilters to work more efficiently because this equipment creates high shear that thins the mud and makes separating the solids easier.

Without crosslinking, an X-C polymer mud is an extremely low-solids fluid with readily controllable viscosity and moderate filtration control. The mud engineer may also add a biopolymer to other water-base muds, where it makes the viscosity high at low shear rates and low at high shear rates to improve the drilling rate. Adding bentonite or other filtration control additive lowers the filtration rate.

Low-Clay-Solids Muds

Oilfield workers often apply the term "low-solids mud" to heavily weighted muds that are actually high in solids due to the addition of barite. As the solids content from barite increases, the viscosity increases and may become too high if the crew does not reduce the amount of clay solids at the same time. So a more appropriate name for these heavyweight muds is *low-clay-solids mud*.

Removing Drilled Solids

Heavyweight muds that contain mainly good-quality bentonite and barite give lower filtration rates and better all-around performance than muds that contain large amounts of poor-quality clays from drilled solids. So, operators often specify that a rig use mud cleaners. Mud cleaners remove low-gravity drilled solids before they break up into fine particles that contaminate the clay fraction (the colloidal phase) of the mud. Then, the crew adds a good-quality bentonite, which rebuilds the clay fraction with just enough viscosity to support the barite and prevent settling. Where contamination by drilled solids is a problem, rigs working in such areas usually have desanders, desilters, mud cleaners, and centrifuges among the surface equipment.

Hole Cleaning

In using heavyweight muds, hole cleaning is seldom a problem since the density of the mud approaches that of the cuttings. Removing the cuttings at the surface allows the mud engineer to keep the yield point and gel strength of the mud low. Low yield point and gel strength minimize pressure drops in the annulus and swabbing and surging when tripping pipe.

Deflocculation

Deflocculation of solids is essential in controlling heavily weighted muds. Nearly all heavyweight muds used in deep, high-pressure wells contain substantial concentrations of lignosulfonates, resulting in effective deflocculation and low filtration rates. However, the excellent deflocculation provided by lignosulfonates requires that the crew closely control the quantity of low-gravity solids in the mud, since the high degree of deflocculation can allow the amount of drilled solids to build up.

Aerated Mud

Another option is *aerated mud*, a fluid that forms when air and mud are injected into the standpipe at the same time. It looks like gas-cut mud. Aerated mud may replace regular drilling mud to prevent lost circulation and increase the penetration rate. One of the causes of lost circulation is drilling mud whose hydrostatic pressure is so high that it fractures the formation. Air or gas in the mud lightens it and reduces the amount of pressure the mud exerts on the formation. Another cause of lost circulation is drilling through coral or cavernous limestone formations. Aerated mud can also help in these cases.

When drilling with air or gas, the operator may switch to aerated mud if water is entering the hole at a rate of more than 50 bbl (8 m^3) per hour.

Some guidelines for drilling with aerated mud are:

1. Do not attempt to aerate just any mud. It must be of good quality and have a low gel strength so that the air can readily break out. Because the mud is recirculated, it must be deaerated after each cycle so that the mud pumps can efficiently pump it. If not deaerated, the air in the mud could gas lock the intake and discharge valves of the pump. Gas locking occurs when air or gas in the mud breaks out of the mud and holds the valves open.
2. Maintain a high pH and use suitable corrosion inhibitors in aerated muds. Because oxygen is in air, and because iron and other metals readily oxidize in the presence of oxygen, corrosion can be very severe, sometimes ruining a string of pipe in a few days. Usually, keeping the mud's pH above 10 and adding suitable inhibitors minimizes or eliminates corrosion in aerated muds.
3. Set enough casing. The annular velocity of the mud may be as high as 2,000 to 3,000 feet (600 to 1,000 metres) per minute in the upper part of the hole, where the air is expanding rapidly. A velocity this high can destroy an open hole unless the formation is hard rock.

Oil-Emulsion Mud

An *oil-emulsion mud* is a water-base mud with a little oil mixed in, usually diesel or synthetic oil. (Synthetic oil is a vegetable or ester oil and is biodegradable. Another synthetic oil is mineral oil, which is less toxic than diesel.) Oil-emulsion muds increase the drilling rate, reduce torque (twisting force on the rotating drill string), reduce sticking of the drill pipe, and increase the life of the bit, all because the oil has a lubricating effect.

When the crew adds oil to a water-base mud, they usually also add an emulsifier. An *emulsion* is a stable combination of two liquids that normally do not mix, like oil and water. Normally, oil and water separate as soon as they stop moving. The addition of an emulsifier keeps everything well suspended (fig. 14). The crew can add up to 5 percent oil into a water-base drilling mud without adding an emulsifier, but even these mixtures are more stable with a small amount of emulsifier. Lignites, lignosulfonate, or soap-type additives are good emulsifying agents.

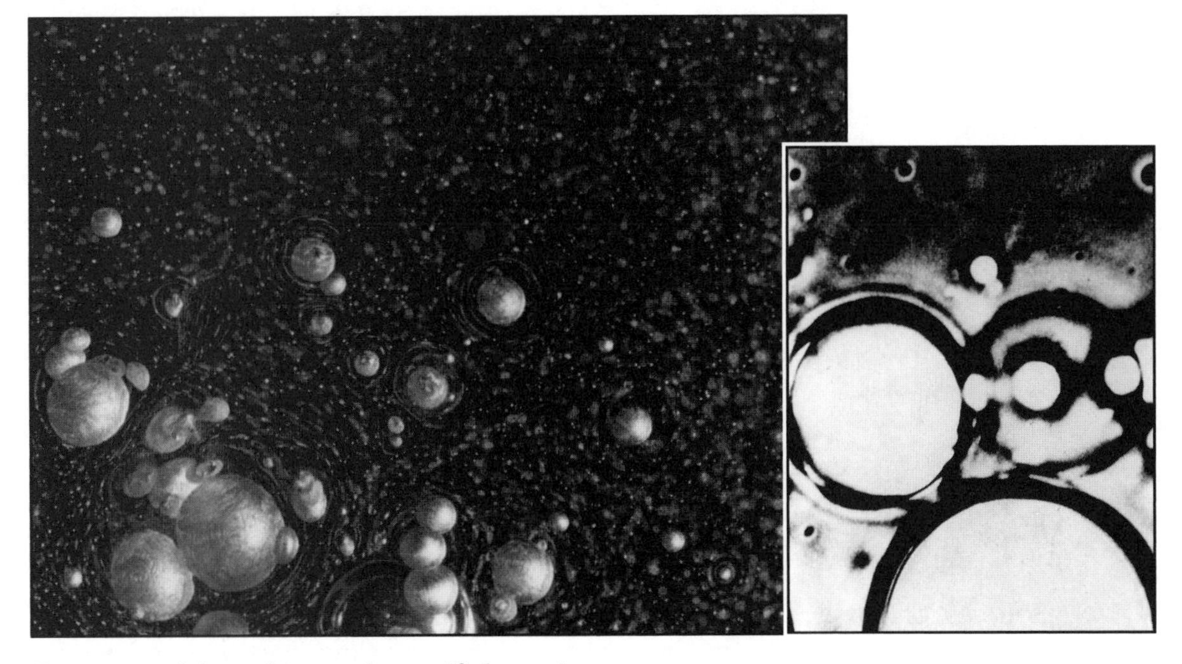

Figure 14. Oil-emulsion mud magnified 900 times

Salt Water

Salt water, or brine, is an alternative to plain fresh water as drilling fluid when the operator wants a fast ROP but the formation contains salt or shale. As mentioned earlier, a freshwater drilling fluid, or a saltwater drilling fluid whose salt content is less than that of the salt water contained in a shale's pore spaces, can cause the shale to swell and slough into the hole. What happens is that the formation salt water, with its high salt content, is attracted to the fresh- or low-saltwater drilling fluid and migrates into the hole. As the salt water moves from the pores into the borehole, it causes the formation to expand and slough into the hole. Also, fresh water or salt water with a low salt concentration can dissolve salt formations, causing hole enlargement and other difficulties.

Salt water is denser than fresh water—its density increases with its salt content to a level of about 10.0 ppg (1,200 kg/m^3) for saturated salt water. For this reason, brine is useful in areas where formation pressure is normal because its hydrostatic pressure is high enough to hold back the formation fluids without adding barite. Saturated salt water (water that has all the salt dissolved in it that it can hold) can drill salt formations without causing serious hole enlargement because it cannot dissolve any more salt. Salt water can also cause less shale damage than fresh water if the salinity of the drilling fluid is the same as the salinity of the connate water in the shale. (Connate water is the water trapped in a formation's pore spaces when the formation was originally laid down.)

The viscosity of salt water increases with salt concentration, and saturated salt water is about 1.7 times more viscous than fresh water at the same temperature. Its viscosity, however, like that of fresh water, may be too low for removing cuttings. Filtration control is also a problem. The mud engineer may use additives to make up a mud with more useful properties.

Sources for Salt Water

Salt water is available from several sources. A nearby oil or gas well often produces salt water that can be transported to a drilling rig, although the crew may need to add more salt at the rig. In offshore drilling, seawater is readily available; again, extra salt may be added. When no naturally occurring sources of salt water are convenient, the crew can mix fresh water with salt at the well site, or the operator can buy saturated brine if well conditions warrant.

Corrosion

One drawback of using salt water over fresh water is that salt promotes corrosion, as anyone knows who has owned a car and lived near the ocean. Saltwater corrosion of steel is worst when the salt content is about 3.5 percent. Adding caustic soda or lime hydrate raises the pH to a range of 9 to 11, which reduces salt water's tendency to cause corrosion.

Salt as a Contaminant

Salt contamination, even in small amounts, thickens freshwater muds and increases filtration rates. (Oddly enough, although small amounts of salt thicken mud, larger concentrations thin it.) Bentonite and other clays hydrate less when added to salt water, so they do not build as much viscosity. These effects lead to special problems in the preparation and maintenance of stable muds containing high concentrations of salt.

Salt Muds

A mud is a saltwater mud, or a salt mud, when it contains more than 1 percent salt (6,000 parts per million of chloride ion) and has not been converted to another type of mud, such as a lime mud or a lignosulfonate mud. The concentration of chloride ions may range up to 189,000 parts per million, at which point it is saturated—that is, no more salt can dissolve in the water. Depending on the amount of salt in the water phase, these muds are referred to as brackish-water muds, seawater muds, or saturated salt muds.

When the available water is brackish or seawater, such as in offshore wells, operators often use it. Salt water is an excellent source of calcium and magnesium, which they can use to control viscosity. Operators spud in offshore wells with seawater mixed with attapulgite and later improve the mud with bentonite, caustic soda, and lignite or lignosulfonate, sometimes adding gypsum for control and stability. Starches control fluid loss.

Controlling Viscosity, Filtration, and Foaming

The saltwater clay attapulgite and asbestos are the usual additives for increasing viscosity in a saltwater mud.

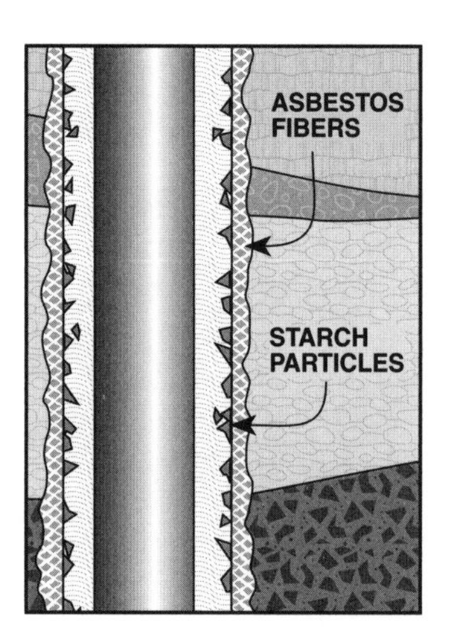

Figure 15. Asbestos fibers form a mat on the wall of the hole that traps other additives and controls fluid loss.

Asbestos

Asbestos fibers of a very fine grade build viscosity in saltwater drilling fluids. The asbestos fibers separate, overlap, and interlock to produce a mechanical gel structure in the water. Relatively small amounts (4 to 8 lb/bbl, or 11.5 to 23 kg/m^3) usually produce a low-solids fluid that thins with high shear rates, permitting fast ROP. The packing of the asbestos fibers produces a loose, porous filter cake and high fluid loss. However, the mat is an excellent bed for additives such as starch, cellulose derivatives, or prehydrated bentonite. These additives combined with asbestos provide good filtration control (fig. 15). Asbestos-based muds tend to settle, or subside, leaving clear water on top, but they are easy to remix by stirring. The advantage of asbestos is that a relatively small amount produces a mud with a good ability to carry cuttings. The disadvantage is that long-term exposure to asbestos without proper personal protective equipment (PPE) has been shown to be carcinogenic. Crew members handling asbestos should therefore wear proper PPE.

Clays

Asbestos and attapulgite develop good suspending properties in salt muds but do not solve the problem of poor filtration. The solution to poor filtration and also to the problem of foaming, which is common in salt muds, is to add higher-grade clays such as bentonite. Prehydrating ordinary clays and bentonites in fresh water before introducing them into a saltwater system increases their effectiveness, as well as lowering filtration. Adding small amounts of lignosulfonate and caustic soda to prehydrated clays also increases the effectiveness of such a mud system.

Cellulose

Mud engineers may also add cellulose derivatives to build viscosity and control fluid loss. Pregelatinized starch is the most common filtration-reducing agent. CMC, useful in low-salt solutions, is not soluble in stronger salt water. Other cellulose derivatives are available that do dissolve in salt water (including saturated salt water), but they are slower to act in saturated salt water. Cellulose derivatives can produce the desired viscosity with low fluid loss alone or in combination with asbestos.

Other Additives

A number of defoamers such as castor oil, pine oil, octyl alcohol, turpentine, and commercial antifoaming compounds combat severe foaming problems. Lignosulfonates are the most common thinners for salt water.

Seawater Muds

Offshore, fresh water is not always available. Since transporting fresh water is inconvenient and expensive, seawater, which contains sodium, calcium, and magnesium, is often the choice for the liquid phase of the drilling mud. Caustic-lignite mud, lime mud, and gyp muds have been used as seawater muds, but, today, lignosulfonate muds have pretty much replaced them.

Two types of lignosulfonate-treated seawater muds are available. In one, added alkaline materials such as caustic soda, lime, or barium hydroxide combine with the calcium and magnesium in the seawater and form a *precipitate*, a material that is heavy enough to easily settle out of the mud. Precipitating the calcium and magnesium requires less lignosulfonate to lower the viscosity and fluid loss and therefore costs less. The pH of this mud is kept at 11 or higher.

In the other type of lignosulfonate mud, the pH is maintained between 9 and 11, which is high enough to allow the lignosulfonate to dissolve in the mud and low enough to keep the magnesium from precipitating. This type of mud sacrifices some mud properties in favor of the shale stabilization that the magnesium provides.

Saturated Salt Muds

Saturated salt muds contain the maximum amount of salt that can be dissolved in a mud. Operators use them for drilling through salt domes and massive salt beds, where a freshwater mud would enlarge the hole.

Preparing a Saturated Salt Mud

There are several ways to prepare a saturated salt mud. To convert a freshwater or lime mud to a saturated salt mud, the crew uses as much of the original mud as possible. With large additions of salt, the mud first thickens, then becomes less viscous, and finally becomes stable. Although the crew adds a lot of water to the original freshwater mud, they can use its solids to increase viscosity and weight, so less saltwater clay is necessary. Usually, the crew discards about half the base mud, saturates what remains, and replenishes the system with saltwater clay and starch. They perform these actions one batch at a time to keep enough mud in the circulating system.

To prepare a saltwater mud from fresh makeup water, the crew adds the salt first. Then they add saltwater clay until the viscosity is where it should be. About 28 lb/bbl (80 kg/m^3) of a good saltwater clay will give a viscosity high enough to suspend barite. Then they mix in starch to control fluid loss through a mixing hopper at a rate of about 10 to 20 minutes per sack so the starch does not form lumps.

Sometimes an operator chooses to saturate a freshwater mud by drilling into salt. This saves the cost of buying salt, but it may cause the hole to enlarge too much and complicate setting the casing. Also, if the crew needs to fish something out of the hole, the top of the fish can get lost in the enlarged section of the hole.

Altering the Mud while Drilling through Salt

The best way to prepare a salt mud to prevent formation salt from dissolving is to mix it in a separate tank (pit) that is isolated from the active tank system. When the new mud is ready, crew members then switch the tank into the active system so the pump can pick up the salt mud and pump it downhole. On small rigs, where a separate tank may not be available, the crew may use a *salt barrel.* A salt barrel is a 55-gal. (200-L) drum with the top cut out and a piece of 2-in. (50-mm) pipe about 3 ft (1 m) long welded upright in the bottom. The top of the pipe is just below the top edge of the drum. A water connection is also welded in the base of the drum. The crew suspends the drum over the mud tank, fills it full of salt, and runs water through it slowly. The water becomes saturated with salt as its level rises inside the drum. When it reaches the top of the drum, it flows into the pipe and out the bottom of the drum into the mud system.

Another way to increase volume or decrease viscosity while drilling salt is to introduce a small stream of water into the return flow line ahead of the shale shaker. The water becomes saturated by dissolving the salt cuttings. The water can also circulate through the salt cuttings below the shale shaker and then into the mud tanks. Sometimes the mud bypasses the shale shaker altogether and salt collects in the mud tanks. The crew then adds water to the tanks and agitates the settled salt cuttings so that they dissolve. This method works only if the cuttings do not include shale or rock that could be recirculated.

Controlling Salinity

The solubility of salt increases slightly as the temperature of the water or mud rises (table 9). Consequently, water that is saturated with salt at surface temperature may dissolve more salt from the formation at bottomhole temperature, because temperature increases with depth. If more salt is dissolved at bottom because of higher temperature, this salt comes out of solution when the fluid returns to the surface and cools. To control the salinity and therefore the amount of hole enlargement, the crew continues to add either water or salt to the active mud system after preparing a saturated salt mud. To guard against hole enlargement, the crew can add an extra 4 to 5 lb of salt for each bbl (12 to 14.5 kg of salt for each m^3) of saturated salt water added to the mud in the tanks. In contrast, they can run a small stream of fresh water near the pump suction to decrease the salinity. Decreasing the salinity brings about some desirable enlargement of the hole when deep drilling in dome salt. Boreholes drilled through salt formations tend to decrease in size because massive salt formations are elastic—that is, after the borehole is drilled into them, the salt swells to decrease the gauge of the hole.

Table 9
Effect of Temperature on Solubility of Salt in Water

Temperature, °F (°C)	Salt to Saturate 1 bbl (0.159 m^3) of Water, lb (kg)
80 (27)	127 (58)
120 (49)	129 (59)
160 (71)	132 (60)
200 (93)	137 (62)

When to Use a Saturated Saltwater Mud

When to use a saturated saltwater mud varies with the conditions downhole. For example, if the casing is set into or just above a massive salt bed or dome, the crew may use saturated salt water alone to drill through the salt. In areas where high fluid losses are acceptable, they may use a saturated saltwater mud with no filtration-reducing additives. But if they encounter a porous zone, the fluid loss causes the filter cake to get too thick, resulting in stuck pipe. Also, as mentioned earlier, drilling through shales with salt water can cause sloughing if the concentration of salt in the fluid is lower than the concentration of salt in the shale's connate water.

To summarize—

Types of muds

- Spud mud—plain water, natural mud, or mud with additives to build wall cake and carry cuttings
- Treated muds include—
  - phosphate to reduce viscosity, gel strength, and filtration rate in uncontaminated natural muds; useful in shallow wells
  - lignosulfonates to deflocculate solids; useful in deep, high-temperature wells
  - calcium (lime) to control shale sloughing, hole enlargement; largely replaced by lignosulfonate
- Low-solids muds for fast drilling rates
- Polymer muds—high viscosity, low fluid loss muds for drilling shale
- Low-clay-solids muds—fast ROP
- Aerated muds—prevent lost circulation and increase ROP
- Oil-emulsion muds—increase ROP, reduce torque and drill stem sticking, and increase bit life
- Salt water or brine—fast ROP in salt or shale formations
- Salt muds—contain more than 1 percent salt; additives include attapulgite, asbestos, or cellulose for increasing viscosity
- Seawater muds
- Saturated saltwater muds

▼
▼
▼

Composition of Oil Muds

▼
▼
▼

Oil muds have oil, usually diesel or synthetic oil, as the liquid phase instead of water. Oil muds are more expensive, harder to handle, and harder to dispose of than water-base muds, but they are simple to prepare and not difficult to maintain. Because of the cost and environmental concerns, operators use them only when the downhole conditions require it. Operators commonly use oil muds for—

1. protecting producing formations,
2. drilling water-soluble formations,
3. drilling deep, high-temperature holes,
4. preventing differential pressure sticking,
5. coring,
6. minimizing gelation and corrosion problems (when used as a packer fluid),
7. helping to salvage casing (when used as a casing pack),
8. mitigating severe drill string corrosion,
9. preventing entrainment of gas, and
10. drilling troublesome shales.

Advantages of Oil Muds

The benefits of oil muds are many, in particular situations. Operators have long used oil muds for coring and completion. Nevertheless, their advantages in difficult formations and deep or directional drilling have made them a more widely used choice in recent years. When water-base muds cause a problem—in formations containing water-sensitive shales, corrosive gases, or water-soluble salts, for example—an oil mud can be the answer. The cost is relatively high, but proper handling, storage, and careful moving of the mud from well to well can make its use cost-effective.

Coring and Completion

Properly formulated oil muds, because they contain little water, do not hydrate clays, dissolve salts in a formation, or react with formation solids. With the right additives, very little of the liquid phase of the mud escapes into the formation as filtration loss. All these characteristics minimize damage to producing formations and minimize the alteration of the rock in a core.

Hole Stability and Corrosion

Oil muds are very useful for maintaining the stability of the hole's walls in shale formations, which tend to slough off in the presence of water. They also help prevent corrosion of steel parts in wells where corrosive gases combine with water in a water-base mud to form acids. They are increasingly popular for protecting the drill string as well as the tubing and casing from corrosion.

Lubrication Ability

The lubricating ability of oil works downhole the same way oil in a gear works—its slipperiness reduces friction. When the drill string or liner gets stuck, an oil mud can come to the rescue if circulation is possible. The slipperiness of oil muds also minimizes torque and drag in directional drilling.

High-Temperature Drilling

Because oil muds also have a higher boiling point than water-base muds, they are very useful in deep wells—for example, in deep water and arctic locations, where high temperatures are the rule. In holes where high temperatures break down high-density water-base muds, oil muds present no problems, even at temperatures over 400°F (204°C). Operators have drilled many 20,000-foot (6,100-metre) holes with oil mud and have experienced comparatively low maintenance costs.

Disadvantages of Oil Muds

The disadvantages of an oil mud are—

1. its expense,
2. the problem of contamination by groundwater (increased water content causes some oil muds to thicken),
3. its flammability,
4. safety and health concerns for the crew (its slipperiness and dirtiness are objectionable and can cause accidents), and
5. usually, a slower drilling rate.

In addition, the operator must take extreme precautions to prevent the mud from contaminating the environment during use, and dispose of it in a safe manner. The two types of oil muds are oil-base and invert-emulsion muds.

Oil-Base Muds

An oil-base mud consists of oil, emulsifiers, stabilizing agents, salt, and less than 5 percent water. Although oil-base mud has a small amount of water, any additional water is a contaminant that must be avoided. Even a very small amount may cause the mud to thicken.

Invert-Emulsion Muds

To overcome the problem of water contamination, researchers looked for more effective emulsifiers to force the water and oil to mix. They succeeded in finding new emulsifiers that led to oil muds in which the water is a useful component instead of a contaminant. Because an oil-emulsion mud is a water-base mud with some oil added, the term *invert-emulsion mud*, or *invert-oil mud*, was coined to refer to oil-base muds with water added. Invert-emulsion muds contain from 10 to 60 percent water in an emulsion. Some of the advantages of an invert-emulsion mud are that any mud with more than 30 percent water does not burn, and it costs less than an oil-base mud. The emulsified water has little effect on the viscosity of the oil.

An invert-emulsion mud is composed of oil, water, emulsifiers, and stabilizing agents. Oil decreases the viscosity of an invert-emulsion mud; water increases it. A disadvantage of invert-emulsion muds is that they have high fluid loss. However, emulsifiers, stabilizing agents, and other special compounds can control it.

A relaxed invert-emulsion mud generally has a higher oil-to-water ratio. Otherwise, its composition is the same as an invert-emulsion mud. Advantages include that the drilling rate is as fast as or better than a water-base mud, contaminants do not affect it, and it is stable at high temperatures.

Additives to Oil Muds

The crew usually mixes oil muds on site. Both oil-base muds and invert-emulsion muds require the addition of an emulsifier, a viscosity increaser, lime, and a weighting material. Other additives work as viscosity reducers and fluid loss control agents.

Emulsifiers

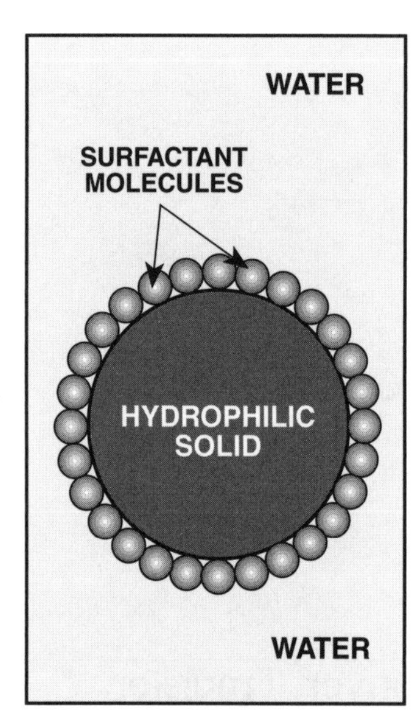

Figure 16. A protective skin of surfactant molecules surrounds a hydrophilic solid.

Most minerals used as mud additives, such as barite, clays, and lime, are *hydrophilic* (literally, water-loving). Hydrophilic additives attract the small amounts of unemulsified water in the mud, which causes the additives to clump and settle out. Hydrophilic additives also make the mud's viscosity and gel strength too high, gum up screens that filter solids out of the mud, coat the drill pipe, and destabilize the emulsion. To combat these effects, the crew adds an emulsifier, or *surfactant*, to oil muds. A surfactant, which is short for surface-active agent, is a chemical that adheres to the surface of a substance. In an oil mud, the surfactant bonds to the hydrophilic solids in the mud (fig. 16). The surfactant keeps water away from the solids so they do not clump together.

By varying the concentration of emulsifier, the mud engineer controls an oil mud's viscosity and gel strength and its ability to suspend cuttings. Some emulsifiers are not effective in highly alkaline or saline water and break down at the high temperatures in deep drilling.

Adding an organic acid and alkali such as lime to the oil forms the emulsifier in an oil-base drilling mud. Lime also combats contamination by carbon dioxide in the air and hydrogen sulfide in petroleum. The lime reacts with these gases to produce water and a harmless salt. Oil muds usually contain lime as a preventive measure, just in case the driller encounters this problem.

One formula for oil-base mud uses a fatty acid such as tall oil, neutralized with silicate of soda. Another contains naphthenic acid, neutralized with lime. A third type of oil-base mud contains a rosin and naphthenic acid mixture, neutralized with lime. Small amounts of other chemicals are often added to serve as stabilizing agents.

Additives to Control Fluid Loss

Lignite combined with a surfactant to disperse it is one way to control fluid loss in an oil mud without using huge amounts of surfactants or asphalt. It is stable at a much higher temperature than lignite in a water-base mud. In the oil-mud environment, contaminants such as salt or high concentrations of calcium do not destroy the effectiveness of the lignite.

The crew may add ammonium humate or asphaltenes instead of lignite for fluid loss control. Asphaltenes form long, sticky molecules that seal extremely small openings in the formations, bringing fluid loss to near zero. They also increase an oil mud's viscosity and wall-building properties.

The disadvantage of asphaltic additives is that, in the presence of emulsifiers, they tend to make a mud with a high viscosity and low gel strength. If the crew adds enough asphalt and emulsifier to develop gel strength, the mud can become extremely viscous, slowing the drilling rate.

Organophilic Clays for Increasing Viscosity

For increasing viscosity, oil muds cannot use bentonite alone because bentonite is hydrophilic. To produce a clay for viscosity control in oil muds, the supplier adds a chemical called an amine salt to ordinary bentonite or other clay. The amine salt converts a hydrophilic clay to an *organophilic clay*, one attracted to organic substances such as oil. Suppliers sell these treated clays under the label of organophilic clays, organic clays, or amine clays. Heavy oil muds that use organophilic clays develop enough gel strength to be good packer fluids to leave in deep, high-temperature wells.

Thinners for Reducing Viscosity

When weighting up an oil mud, the crew may also add a thinner to prevent it from becoming too viscous. In some systems, the thinner lowers the viscosity of the oil phase. In others, the thinner is a surfactant. In either case, the danger is that too much thinner can weaken the emulsion.

Another type of thinner is an organic sulfonate. Relying on organic sulfonates is also tricky. They work by stripping the amines from treated organophilic clays, so the crew could ruin the mud by adding too much sulfonate and then compensating by adding more clay, and so on. In addition, if fine drilled solids have increased the mud's viscosity, adding sulfonates can increase the viscosity even further.

Weighting Materials

Barite is the main material added to weight up oil muds, as it is for water-base muds. Hematite is another popular material because it allows greater weight control while maintaining a lower solids content. A few oil muds contain calcium carbonate (either limestone or crushed oyster shells) as the weighting material, which helps prevent lost circulation as well.

Synthetic Muds

Operators and the rig crew members must be aware of the several effects that drilling mud can have on the environment. Drilling muds can contain components that are toxic to human, animal, and plant life. Oil muds are especially damaging. Unseen effects of mud, for example, include pollution of groundwater, which may be a drinking water supply. Offshore, release of mud can damage marine life. Visible effects include pollution of surface water and land, affecting soil productivity. The U.S. and other governments have enacted legislation to protect the environment, and rig personnel must be familiar with the laws in order to comply.

At the same time, the search for petroleum has resulted in drilling deeper wells in increasingly difficult environments, such as in deep water and arctic locations, which has presented problems such as high downhole pressures and temperatures. Traditional oil muds that can handle these adverse conditions are also the most environmentally hazardous.

Therefore, engineers developed synthetic muds that have a lower impact on the environment and are easier to dispose of safely. For example, some new invert-emulsion muds are based on vegetable oil or ester oil, which are biodegradable. Others based on mineral oil have lower toxicity than ordinary oil muds.

Mud companies have also developed environmentally safe components for water-base muds. Such additives include PAL (polyanionic lignin), polymers such as VA/VSC (vinyl amide/vinyl sulfonate copolymer) and MPT (modified polyacrylate terpolymer), and an oxygen scavenger to control corrosion. (Corrosion can be a big problem at high temperatures.) Using environmentally safe muds lowers the normally high cost of disposing of cuttings, which are coated with the mud. In addition, programmers have developed computer software that helps operators understand and monitor their compliance with environmental regulations.

To summarize—

Types of oil muds

- Oil-base muds—contain less than 5 percent water
- Invert-emulsion muds—contain 10–60 percent water
- Synthetic muds based on vegetable or mineral oil—less environmentally hazardous

Additives to oil muds

- Emulsifiers—prevent hydrophilic additives, such as barite and clays, from clumping and settling out
- Fluid loss additives
  - Lignite
  - Ammonium humate
  - Asphaltenes
- Viscosity increasers—organophilic clays
- Viscosity reducers
  - Surfactants
  - Organic sulfonates
- Weighting materials
  - Barite
  - Hematite

▼
▼
▼

Air, Gas, and Mist Drilling

Although operators don't use air or gas very often, it's a valuable method of drilling when the formation allows it.

Advantages

Penetration rates with air and gas are higher—partly because air or gas cleans the bottom of the hole more efficiently than mud. Mud is denser than air or gas and tends to hold the cuttings on the bottom of the hole. As a result, the bit cannot make hole as efficiently because the bit redrills some of the old cuttings instead of being constantly exposed to fresh, undrilled formation. With air or gas as a drilling fluid, the cuttings literally explode from beneath the bit cutters.

Air or gas drilling also gets more wear from the bit because air and gas are much less abrasive to the metal parts than drilling mud. Both air and gas do an excellent job of cooling, and both transport cuttings to the surface quickly. In addition, with air or gas circulation, the formation is easy to identify, and it is easy to detect the presence of gas, oil, or water.

Disadvantages

Unfortunately, air or gas drilling has several disadvantages that overshadow the advantages. First, if the formation rock is soft and the walls of the well tend to slough into the hole, air or gas does not have enough hydrostatic pressure to prevent them from doing so. Sloughing walls could cause the drill stem to stick. Second, preventing formation fluids from entering the wellbore is impossible because neither air nor gas can exert enough pressure to keep them out. This second disadvantage is especially important because most wells encounter water-bearing formations at some time.

The water easily enters the borehole and, if the quantities are large enough, makes the fine cuttings ball up and clog the hole. The low hydrostatic pressure also creates the hazard of a kick if the bit drills into a high-pressure formation.

Another disadvantage is that the hazard of fire or explosion is always present. Natural gas is flammable on its own, and air can mix with formation gas and become flammable. Also, corrosion of the drill stem can be a problem—corrosion is the reaction of the oxygen in air with metal. Although chemicals to combat corrosion are available, the operator must consider the added cost and effort of using them.

Dry Air Drilling

Dry air drilling, or *dusting*, involves the injection of dry air or gas into the standpipe at a high enough rate to move through the annulus at 2,000 to 3,000 feet (600 to 900 metres) per minute. To use dry air drilling, the formation must be either completely dry or so nearly dry that the air can absorb the moisture and the cuttings leave the blooey line in a cloud of dust (fig. 17). Since air imposes very little pressure on the formation, dry air drilling usually makes for a fast penetration rate with few hole problems. Dusting does have some problems particular to it. The two main problems are not enough air and too much water.

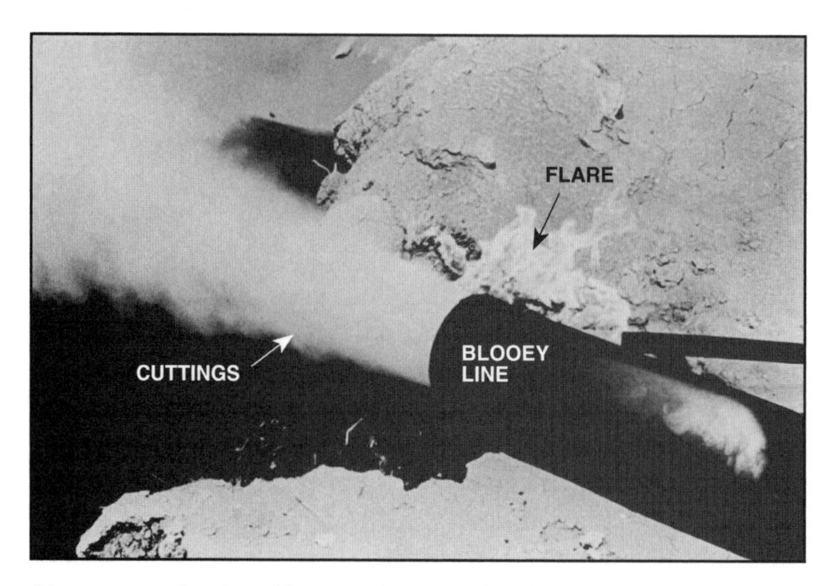

Figure 17. Cuttings blown to dust exit the blooey line.

Insufficient Air or Gas

To circulate a gaseous fluid such as air or natural gas requires a large volume of fluid at a high pressure and a fast velocity. Most of the problems encountered in dry air drilling are caused by the volume of air being too small to create the velocity needed to carry the cuttings up the annulus. A volume of air too small may occur because the capacity of the compressors is too low or because the hole has become enlarged. Since the deeper the hole, the more air is needed, it is quite common for the compressor capacity to be adequate at a depth of around 4,000 feet (1,200 metres) but inadequate at greater depths of around 6,000 feet (1,800 metres). Sometimes, an operator attributes drilling difficulties to hole enlargement, incompetent formations, or water in the hole when the problem is actually not enough air. (An *incompetent formation* is one composed of materials that are not bound together, like sand.)

On the other hand, too much air can cause the hole to enlarge in certain soft, unstable formations. Drilling experience and careful observation of the drilling process are the only ways to determine why a hole has enlarged.

Water in the Formation

Hole problems often begin when the formation being drilled produces enough water to dampen the cuttings, causing them to ball up. Usually the first indication of this problem is a lack of dust at the blooey line. What happens downhole is that the cuttings form a ring of mud that can cause the drill string to stick. A slow increase in standpipe pressure is the signal that a mud ring has formed and packed off the hole.

The usual way to remove a mud ring is to inject a few gallons (litres) of detergent foamer and then raise and lower the drill string, passing the tool joints through the ring a few times. Then the crew must dry the hole before dusting can continue. Simply circulating warm, dry air for some time may be enough to dry the hole, or injecting a drying agent such as silica gel or other commercial agent continuously may be necessary. But it is usually not economically feasible to dry up more than 2 to 3 bbl (0.3 to 0.5 m^3) of water per hour. If the formation continues to flow more than this quantity, conversion to mist drilling may be the best solution.

Mist Drilling

When an air or gas drilling operation encounters a formation containing water, the water flows into the wellbore and begins to fill it. Water in the wellbore causes two problems. First, because water is heavier than a gas, lifting it out requires more and more gas pressure. Eventually, so much water gets into the hole that the available pressure cannot overcome the weight of the water. Second, the wet cuttings can stick together, fill the annulus, and shut off the flow of air or gas to the surface. The wet cuttings also stick to the bit and drill string. Stuck pipe is the result.

If the quantity of water entering the hole is no more than about 50 bbl (8 m^3) per hour, the solution is to add a foaming agent mixed with water to the air or gas stream. Adding a foaming agent creates *mist drilling* or *foam drilling*. The foaming agent is a chemical similar to soap that causes the water in the annulus to froth and foam into a larger volume. Foam is lighter than water because it is full of air, so less pressure can move the water out of the hole. As mentioned earlier, mist drilling can handle as much as 50 bbl (8 m^3) of water per hour entering the hole. Larger amounts of water require switching to aerated mud or ordinary water-base mud.

Foam keeps the cuttings separated and helps remove liquid from the hole. In mist drilling, mist and bits of foam are circulated out through the blooey line. In mist drilling, a dependable injection pump for adding the foaming agent is extremely important. Should the pump fail, drilling usually must stop to repair or replace it.

Air and Water Proportions

Use as little water as possible but enough to keep the cuttings separated and to keep the blooey line discharging at a steady rate. The amount of injection water to use varies from 6 to 12 bbl (1 to 2 m^3) per hour. The amount of foamer varies from 1 to 2 qt (0.9 to 1.8 L) per hour up to 2 to 4 gal (7.5 to 15 L) per hour, depending on the type and amount of liquid the well is producing. For example, when the liquid is salt water or oil, which act as defoamers, add more foamer or change to one that works in salt water or oil.

Mist drilling usually requires at least 30 to 35 percent more air than ordinary air drilling because of the extra weight of the liquid in the airstream. Insufficient air can become a problem if the formation suddenly starts producing extra water.

Mud Misting

In water-sensitive formations, mist drilling can cause the walls to slough, which enlarges the hole and in turn increases the volume of air needed. The problem of sloughing can sometimes be offset by resorting to *mud misting*. In mud misting, a thin mud slurry instead of clear water carries the foamer (fig. 18). The mud slurry seems to solve the problem by coating the wall of the hole with a thin protective film and providing a top-quality foam that removes the cuttings and any well fluids.

The mud used for mud misting is a special mixture of about 10 lb/bbl (28.5 kg/m³) bentonite, 1 pound per barrel (2.9 kg/m³) soda ash, and ¼ to ½ lb/bbl (0.7 to 1.4 kg/m³) high-viscosity CMC.

Figure 18. In mud misting, a mist of mud and foam carries cuttings out the blooey line.

Watching the Blooey Line

The blooey line is a useful indicator of what is happening downhole and the crew should keep a close eye on it. For example, if no cuttings are blowing out or cavings start to appear, the hole may be sloughing. A drop in the amount of foam in the blooey line stream indicates that drilling has hit another water-bearing formation. The crew may need to add more foaming agent to take care of the increased water flow.

To summarize—

Advantages of air, gas, and mist drilling

- Fastest drilling rates
- Greatest wear from bits
- Easy identification of downhole conditions

Disadvantages of air, gas, and mist drilling

- Low hydrostatic pressure that cannot prevent sloughing and caving of the walls, formation fluids entering the hole, and kicks
- Danger of fire or explosion
- Increased corrosion

Dry drilling or dusting

- Requires a large volume of air or gas at a high pressure and a fast velocity
- Fluid in the hole can create a mud ring that causes drill pipe to stick

Mist drilling

- Contains a foaming agent to lighten and lift water downhole
- In mud misting, a thin mud slurry instead of clear water carries the foamer

Drilling Fluid Problems

▼
▼
▼

Every well presents its own problems to drilling. Problem sources include the type of rock that makes up the formation, the pressures and temperatures in the well, and contaminants that affect the fluid. The mud engineer tailors the drilling program to each well to get the petroleum out in the most efficient manner, at the least cost, and while maintaining control of formation pressures.

In trying to achieve the best performance for all the functions a drilling fluid must perform, the mud engineer unfortunately comes across problems. For example, weighting up a mud to achieve the best transport of cuttings carries the risk of weighting it up so much that it fractures the formation. Drilling experts are always balancing the advantages of making a change in a drilling fluid against the problems that such a change could cause.

Drilling in Shale Formations

Shale is a porous rock but it has virtually no permeability. Frequently, however, salt water and other fluids such as hydrocarbons are contained in the pore spaces. The salt water is connate water, water that existed in the formation when it was formed in the ancient past. In a few cases, hydrocarbons also exist in impermeable shale. One example is in the western U.S., where vast shale deposits hold hydrocarbons that cannot flow into a well. Eventually, these deposits will be mined or extracted in some other way when the need arises. In any event, when a wellbore exposes shale to drilling fluid and where salt water is contained in the shale, the drilling fluid usually requires special attention.

Shale Composition

Shales are sedimentary rocks formed under ancient oceans. They are composed mainly of compacted beds of clays and silts. In soft, or unconsolidated, form, geologists may call them muds or clays, depending on their water content. In more consolidated form, they are termed shales and argillites. Even harder forms are slate, phyllite, and mica schist. The deeper a shale is, the denser and harder it is because the overlying rock, or *overburden*, compacts it (fig. 19).

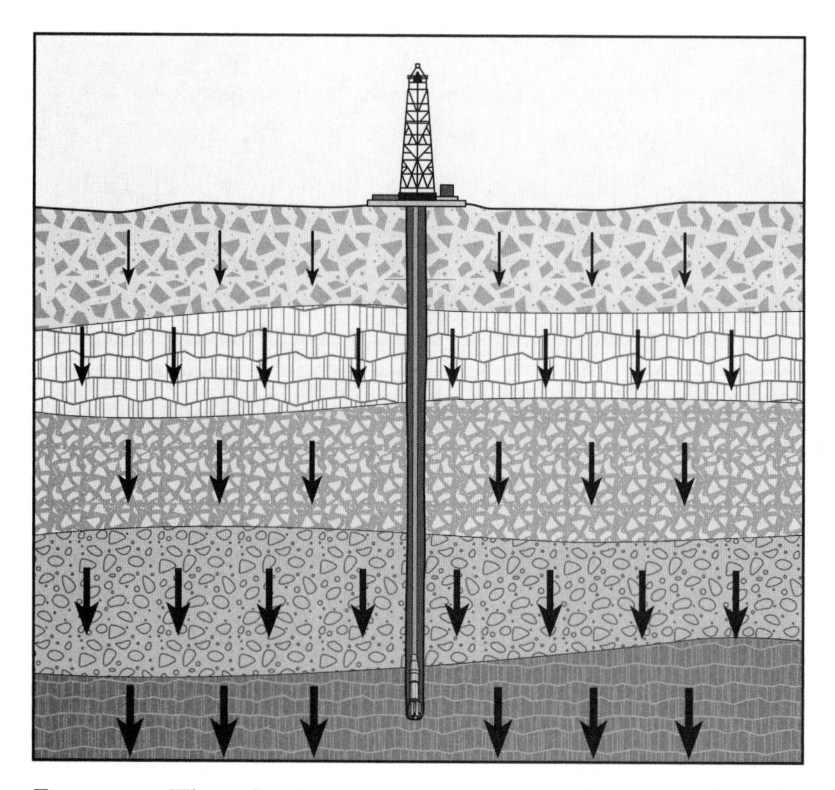

Figure 19. The rocks above put pressure (overburden pressure) on the rocks below and compact them.

Causes of Shale Instability

Shale formations can be unstable, in the sense that, when exposed by a borehole, they create problems, such as sloughing. Causes of shale instability include abnormal formation pressure, overburden pressure, geologic stresses, or water absorption.

Abnormal Formation Pressure

Formation pressure is the pressure that the fluids in the rock exert. Rock fluids exert pressure on the borehole when the borehole is drilled into the rock. Formation pressure may be normal or abnormal.

The oil industry defines normal formation pressure as the hydrostatic pressure that a column of salt water exerts. So, if the pressure of the fluids in a rock equals the hydrostatic pressure of a column of salt water of the same depth, oil people call it *normal formation pressure*.

If for some reason formation pressure is higher than that which a column of salt water exerts, formation pressure is abnormally high. Abnormally high-pressure formations are sometimes referred to as geopressured formations, or, because shale often contains fluids under higher-than-normal pressure, geopressured shales. Geopressured shales are a problem if the mud weight is not high enough to balance the pressure contained in them. A large difference in pressure between the pressure in the formation and the drilling fluid in the hole can cause the shale to *spall* (break off in chips or scales) and fall into the hole. These chips can degrade the mud, cause pipe to get stuck, and enlarge the hole.

If formation pressure is lower than the pressure expected for a column of salt water to exert at a given depth, it is abnormally low, or subnormal. Subnormal pressure can occur in partially or totally depleted formations, formations at high elevations, and formations that outcrop downhill from a well.

Overburden Pressure

Overburden pressure is the pressure that the overlying rocks exert on a particular formation. When a subsurface layer of rock is not connected to the surface in any way, the pressure of the fluid in the pores (the formation pressure) is not relieved until the bit drills into it. A geopressured shale is one such layer. The combination of formation pressure and overburden pressure can cause shales to behave like a plastic fluid and squeeze into the hole.

Geologic Stresses

Movements of the crust of the earth, such as folding and faulting, can also cause shale instability. These kinds of movements are so huge that they produce great stresses in the rock that is folding or faulting. Some shales can deform as the earth moves, and the stresses are absorbed and relieved. But other shales are brittle, and these tend to hold the stresses until drilling releases them.

Regional geology often contributes to shale problems. The Atoka, Springer, and Morrow formations of the mid-North American continent are good examples of troublesome geological alignments. These shales often incline considerably from the horizontal, and they tend to fall into the wellbore when drilled. This problem can be worse if they become wet with water or oil.

Water Absorption

Absorption of water from drilling mud into a shale only aggravates problems that may already make the shale unstable. Fluid loss from water-base drilling muds can cause the shale to swell, shale particles to disperse into the water, and water to enter the rock along cleavage or fracture lines, which results in sloughing (fig. 20). Sloughing does not always happen; sometimes the mud has no detrimental effects on the competence of the shale formation. Brittle shales tend not to swell, but water wets any existing fracture surfaces, cleavage planes, and partings. This causes the hole to enlarge due to the water eroding the rock or the action of the drill string rubbing against the side of the hole.

Shale cuttings that disintegrate while moving from the bit to the surface and water-sensitive sections of shale can disperse into the drilling mud as fine particles. This entry of shale particles to the drilling fluid is called *mud-making*—they mix with the water and make mud. Some mud-making during shale drilling is normal. It is a problem only when the crew is adding so much water to dilute the drilling fluid that it indicates that the wellbore has deteriorated. Drilling mud affects even shales that do not tend to hydrate to some degree, and under certain conditions can cause severe wellbore instability.

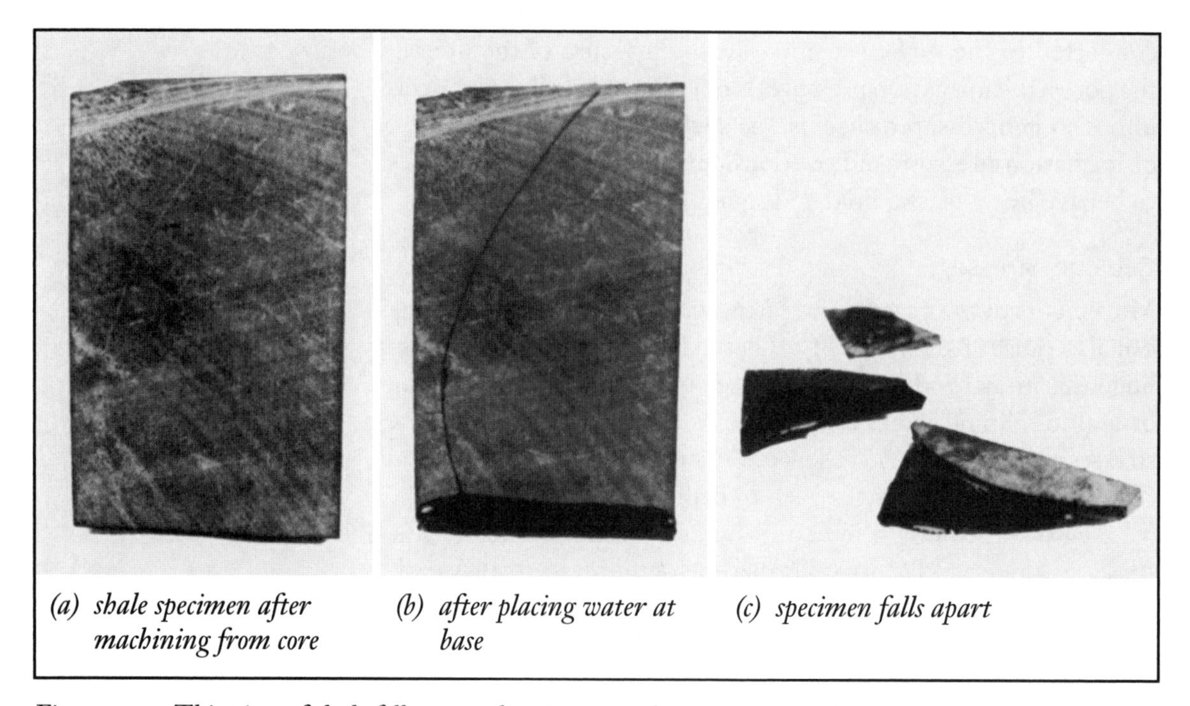

(a) shale specimen after machining from core *(b) after placing water at base* *(c) specimen falls apart*

Figure 20. This piece of shale fell apart when just one edge was wetted with water.

Differential Pressure Sticking

Differential pressure is probably the reason most pipe gets stuck in the hole. Differential pressure is the difference between two fluid pressures; for example, the difference between the hydrostatic pressure of the mud and the pressure in the formation. Normally, hydrostatic pressure is greater than formation pressure. This pressure differential causes pipe to stick because it tends to force the pipe into the filter cake when the drill pipe touches the wall of the hole. The higher the mud weight is, the greater the differential pressure usually is and the harder the pipe sticks.

Muds with high fluid loss worsen differential sticking. High fluid loss creates a thick wall cake and, when combined with higher pressure in the hole holding the drill stem firmly in the wall cake, getting the pipe unstuck can be more difficult.

Varying Mud Weight in the Hole

It is common to think of the mud weight as being constant throughout its circulation path. Actually, mud weight is usually not uniform throughout the circulating system. In general, mud is lightest as it enters the hole through the drill string. When it is returning back up the annulus, it is loaded with cuttings and possibly sand and silt from drilling. These solids add to the mud weight until the shale shaker, desander, and desilter remove them on the surface. It is in the annulus, of course, where differential sticking occurs.

This normal weight increase in the annulus can be a serious problem when drilling is fast in a large hole—fast drilling with a large bit creates more cuttings than slower drilling in a smaller hole. In such a case, the mud weight in the annulus above the bit may be several ppg (kg/m^3) higher than the weight of the mud when it first entered the drill string. For example, a mud that weighs 9.5 ppg (1,138 kg/m^3) at the surface may weigh as much as 12.2 ppg (1,462 kg/m^3) in the annulus above the bit. A 12.2-ppg (1,462-kg/m^3) mud in the annulus rather than the 9.5-ppg (1,138 kg/m^3) mud the crew expects makes differential pressure sticking and lost circulation much more likely.

Reducing Mud Weight to Prevent Differential Sticking

The crew can reduce weight increases in the annulus, if necessary, by adding water to the mud, increasing the circulation rate or reducing the drilling rate. Adding water may not be the best solution when using a highly treated mud because the crew would then have to add more of everything else as well.

Lost Circulation

Lost circulation refers to whole mud that is lost to a formation—all or part of the circulating mud fails to return to the surface. Another term for lost circulation is *lost returns*. Where does the mud go? Either into faults or fractures in the formation or into gravel or cavernous beds that drilling exposed in the hole. The fractures can be natural or they can be caused by a mud weight that is so heavy that it fractures the formation.

Do not confuse lost circulation with reductions in the volume of mud due to fluid loss. In fluid loss, just the liquid phase flows into the formation. For lost circulation to occur, there must be openings in the formation large enough for whole mud to enter and enough pressure to force the mud through these openings (see figs. 2 and 7).

Lost circulation is one of the most serious and expensive problems in drilling. Lost circulation problems are not confined to any one area; they may occur at any depth, regardless of whether the drilling mud is weighted or not.

The types of formation where lost circulation is a problem are—

1. formations containing caverns or open fissures,
2. very coarse and permeable shallow formations such as loose gravel,
3. naturally fractured formations, and
4. easily fractured formations.

Another cause of lost circulation is poorly cemented casing. In this case, the mud can flow into the void spaces left at the casing shoe by a lack of cement in the area. In cases where the mud weight has been increased to drill the formations below the casing, the heavy mud can fracture the formation behind the casing near the shoe and be lost to these fractures. The crew can often fix poor cement jobs by remedial cementing techniques, but it is considerably less trouble and less expensive to ensure that the original cement job was performed properly.

Cavernous and Open-Fissured Formations

Mud losses to cavernous and open-fissured formations generally happen when drilling limestone reefs. Such mud losses are fairly predictable because the formations are easy to trace. The correlation of mud losses in these formations with the extent of the formation is so exact that the driller can see its outline, both its area and depth. Generally, these caverns are below the water table and are full of liquid under pressures that are normal for existing geological conditions. Lost circulation may occur very suddenly, and at the same time the drill string may drop sharply as it decompresses. This rapid loss of mud is a result of the openness of the trouble zones rather than subnormal pressure conditions.

The lowering of the mud level can allow the exposed upper formations to kick because the hydrostatic pressure of the mud column drops and may not be enough to control the fluid pressure in the upper formations. The mud level drop can also cause formations to slough and drill pipe to stick.

When drilling cavernous or open formations, keep the pump suction high so that all the mud will not be immediately pumped away. A drop in mud level is rarely so sudden that a kick or stuck pipe occurs immediately. Even after circulation has been lost, catastrophes can usually be prevented by constantly pumping a small stream of mud into the annulus.

Drilling Blind

Usually, it is impossible to seal or plug a cavern with cement: unless the cavernous openings are small or the cavern itself is small, cement simply flows into the caverns and never fills them. Although drilling without circulation, or *drilling blind*, is not normally a good idea, it is generally the only solution when circulation has been lost to a cavernous formation.

Drilling blind involves pumping water down the drill pipe to cool the bit and carry the cuttings into the cavern. The water does not return to the surface, of course. After drilling a few feet (metres) into solid formation below the cavern, the crew runs casing and cements it at the shoe. They also cement the casing just above the problem zone.

Floating Mud Column

If the formations above the cavern that might kick or slough are exposed, the crew can use a floating mud column, sometimes called a mud cap (fig. 21). Creating a mud cap involves filling the hole with mud until the hydrostatic pressure is only slightly greater than the fluid pressure in the trouble zone. The crew then pumps water down the drill pipe and up the annulus. When it reaches the cavities, the water and the cuttings will enter them, leaving a static mud cap in the annulus above the cavernous formation.

No mud returns to the surface unless the pumps put out more water than the cavities can hold. Sometimes the crew pumps fluid both into the annulus and down the drill string. During trips, the crew must pump some mud into the annulus to maintain a pressure overbalance. Operators have used floating mud columns in the cavernous limestones of Saudi Arabia and southern Florida and in fractured limestones around the world.

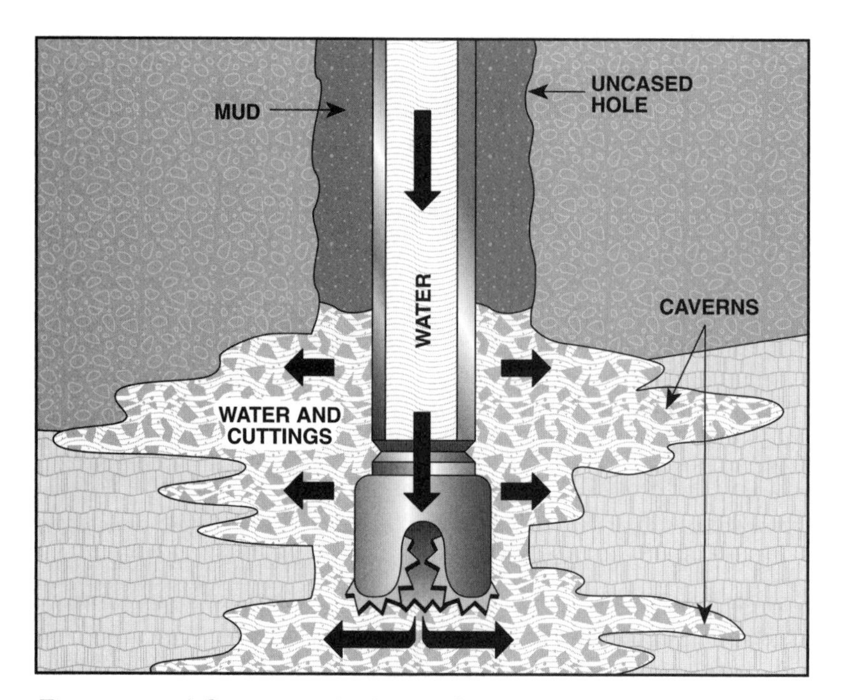

Figure 21. A floating mud column "floats" on a bed of water above the cavernous formations.

Coarse-Grained Permeable Formations

Coarse-grained formations such as sand and pea gravel vary widely in degree of permeability. Some of these formations take on mud and others do not. Whether lost circulation is likely in a particular formation depends on the size of the pores in the formation relative to the size of the particles in the mud. As a general rule, the diameter of the formation pores must be about three times larger than the diameter of the largest particle occurring in quantity in the mud (fig. 22). Formations of this type often have subnormal formation pressure, which also makes lost circulation more likely.

When drilling coarse-grained formations, reducing mud weight as much as possible helps prevent lost circulation. Ways to lower mud weight include—

1. introducing a high concentration of oil into the system;
2. using a fine-mesh shale shaker screen to remove drilled solids; and
3. using desanding equipment if sand is present in the system.

If mud weight is at a minimum and lost circulation is still a problem, thickening the mud to slow the rate of loss by adding a few sacks of lime or cement may be effective. But only try this method when drilling in shallow, low-pressure formations.

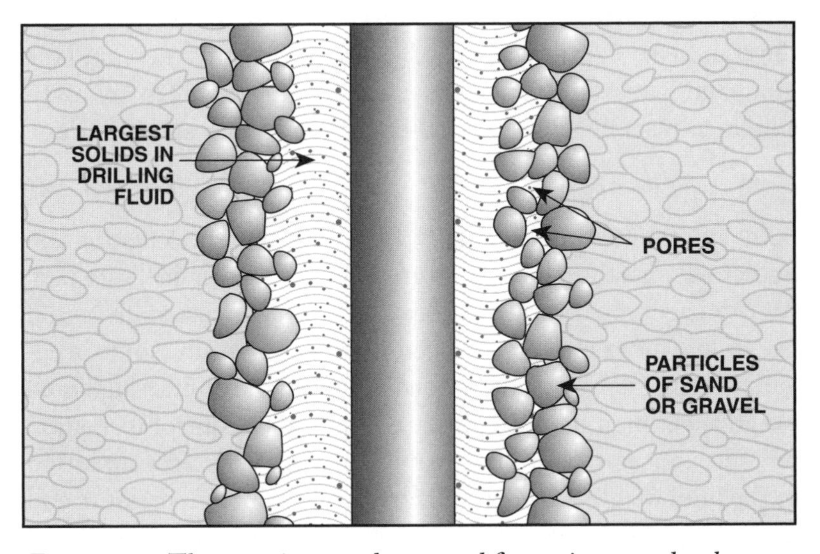

Figure 22. The pores in a sand or gravel formation must be about 3 times as large as the largest particles in the drilling mud for lost circulation to occur.

Naturally Fractured Formations

When the bit drops suddenly, many drillers assume that they have drilled into a cavern; however, the bit may have entered a *fracture zone*. Lost circulation can occur in formations with natural fissures or fractures. In many cases, natural fractures that would be impermeable under normal conditions open and take in mud when the pressure of the mud exceeds some critical point.

Preventing Lost Circulation

Once imposed pressure has opened the fracture, mud rapidly flowing into it can actually widen the fracture. Even after the pressure drops, the opening may not close completely, allowing circulation loss to continue. One way to combat this problem is to maintain mud weight at a minimum density. Sometimes using a thin mud or clear water helps prevent fractures from widening, but lost circulation may continue regardless of preventive measures. If circulation loss is inevitable, the operator may use a cheaper fluid to keep costs down.

In an attempt to prevent lost circulation, some operators add 2 to 12 lb/bbl (6 to 35 kg/m^3) of bulk materials to the mud before reaching an expected lost circulation zone. This procedure is most successful when using mud that is low in solids and has a low viscosity. Pretreating a high-solids mud increases bottomhole pressure losses (thus increasing bottomhole circulating pressure in the annulus) because of increased viscosity, cake thickness, and the mud's tendency to ball up. Consequently, adding bulk materials to a high-solids mud can cause rather than prevent lost circulation.

Easily Fractured Formations

Some formations are particularly vulnerable to lost circulation and operators must therefore take a great deal of care in drilling them. Abnormally pressured formations on the Gulf Coast of Texas and Louisiana, for example, have been particularly troublesome in regard to lost circulation. These formations fracture under high mud weights, yet high mud weights are required to control formation pressure. Successful drilling requires paying very close attention to the mud weight so that it only balances formation pressure and does not overbalance it. Excessive downhole pressure because of high mud weight, pressure surges, and blockage of the annulus is the main cause of lost circulation in these cases.

High Downhole Pressure

An important, yet preventable, cause of lost circulation is pressure surges from running in drill pipe too fast. This practice can create pressures of more than 2,000 psi (13,800 kPa) greater than the hydrostatic pressure of the mud column. You can imagine how surging works by thinking of shoving an empty straw quickly into a thick milkshake. If you watched the top surface of the shake, you would see it pulse upward and then relax back down as the shake enters the straw from the bottom.

Mud weight also accounts for increased bottomhole pressure—the heavier the mud, the greater the pressure. For example, an increase in mud weight of 1 ppg (120 kg/m^3) increases the bottomhole pressure 520 psi (3,588 kPa) at a depth of 10,000 feet (3,050 metres).

Another cause of excess downhole pressure buildup is any closure or reduction of the annular clearance space. Accumulated cuttings or sloughed-off shales may completely close off the annulus. When this happens, pressure from the pump is then added to the hydrostatic pressure. If the formations are strong enough to withstand this additional pressure, the extra pressure may move the blockage up the wellbore. If the formations are weak, however, the pressure may create fractures and force the mud into them. This condition usually results in stuck drill pipe.

Methods of Preventing Fractures

To prevent fracturing and loss of mud returns, take the following precautionary measures.

1. Rotate the drill pipe before starting the pump. This motion begins ungelling the mud and therefore allows a lower pump pressure to break circulation.
2. Break circulation slowly. Circulate at a slow rate and low pressure until good returns are obtained without losing mud; then increase the circulation rate to the required value.
3. When running the drill string back into the hole, stop and restart circulation at one or more points for a period of 5 to 10 min before reaching total depth. Circulating for short periods starts the ungelling process and allows a lower pressure to break circulation at total depth.
4. Maintain the minimum mud weight necessary to control the well.

5. Maintain a minimum annular velocity. An annular velocity ranging from 80 to 120 ft/min (25 to 35 m/min) is usually enough to clean the hole. Higher annular velocities increase the annular pressure losses and impose greater hydrostatic pressure on uncased formations.
6. Keep the mud viscosity and gel strength to the minimum necessary to prevent settling of weighting materials. Low viscosity and gel strength result in lower annular pressure losses and thus lower hydrostatic pressure on the uncased formations.
7. Lower the drill string slowly while tripping into the hole; otherwise, the string may impose severe pressure surges on the hole. The amount of the pressure surge depends on mud properties, hole size, depth, drill pipe and drill collar size, bottomhole assembly diameter and length, and the number of casing protectors installed on the drill pipe. When running casing into the hole, it is very important to lower it slowly because the size of the casing and the effect of the casing float shoe can contribute to pressure surges.
8. When lost circulation occurs, watch the mud level in the hole carefully for a gain or loss of mud. A drop in the mud level can reduce the hydrostatic head on the formation to such an extent that a kick can occur. If protective casing has been set, pull the drill string up into the casing when lost circulation starts. This action prevents the pipe from sticking and positions the drill string properly for an initial attempt to regain circulation after replenishing the volume of mud.

Methods of Limiting Lost Circulation

If prevention has failed, continuing to pump large volumes of mud is usually futile without taking some remedial steps. Some methods to limit loss of mud follow.

1. Reduce mud weight.
2. Pull the drill string into the casing, and stop circulation for six to eight hours, carefully controlling drilling operations when they are resumed. With this procedure, the crew can sometimes restore circulation without altering mud properties and they may even be able to increase mud weight by 1.2 to 1.3 ppg (144 to 156 kg/m^3) without adding plugging materials.

3. Spot soft plugs, which are batches of mud thickened with bulk materials or low-strength cement.
4. Spot mud containing a high concentration of bridging materials, or lost circulation materials (LCMs). LCMs are fibrous, flaky, or granular materials that help seal the formation.
5. Use special drilling methods such as blind drilling, drilling under pressure, and drilling with air or aerated mud.

After circulation in a well has been lost and then restored, the formations that previously took mud seem to become stronger as drilling proceeds, and the operator can use higher mud weights. This strengthening of the formation makes it possible to drill ahead without further loss.

Adding Lost Circulation Materials

As previously stated, the crew can add a material to the mud that seals fractures by plugging them. Lost circulation materials come in a variety of shapes and sizes: they consist of grains, flakes, or fibers and can be coarse, medium, or fine. LCMs include—

- short, weak fibers such as leather, paper, pulp, wood, and cane;
- longer, stronger fibers such as hemp, flax, or synthetic fibers;
- flakes of cork, mica, cellophane, polyester, or plastic (in sizes from ⅛ to 1 in., or 3 to 25 mm, long);
- heat-expanded minerals such as flue ash, volcanic ash, or perlite; and
- granular materials with angular edges such as crushed rock, ground plastics, and walnut or almond hulls.

When using bulk materials to correct lost circulation, the batch method is more effective and less expensive than treating the entire mud system. Also, varying particle size is more effective than using a single particle size. To get such a variety, mix several types of LCMs. The volume of LCM added to each bbl (m^3) of mud will vary considerably, but 50 lb/bbl (about 150 kg/m^3) is not uncommon. Chopped cellophane and ground walnut hulls are one effective mixture.

Lost circulation material fine enough to pass through the shaker screen is too costly to be added as a preventive measure. In some areas, operators add coarse material at predetermined depths, but this procedure is also costly and may actually promote rather than prevent lost returns. When using coarse material, the mud must bypass the shaker. Coarse materials increase mud viscosity, reduce the drilling rate, and increase the pressure drop, all of which may cause more trouble in the hole.

The choice of which LCM to use in a particular well depends on experience in the area. Although reports from offset wells may be helpful in selecting the most effective LCM, do not assume that circulation will necessarily be lost in a particular well at the same depth and mud weight as in an adjoining well. The operator must vary the materials and methods of limiting lost circulation because a remedy that works once in a particular well may not be effective in a nearby well.

Cementing, Plugging, and Thickening

When sealing materials do not correct lost returns within a reasonable length of time, cementing is a possibility. Modified cement compositions are useful with lightweight muds, and neat cement with higher weight muds. Usually, the crew also adds 0.5 to 1 lb (0.23 to 0.45 kg) of LCM per sack of cement.

Thickening and cementing compounds include thickened oil-based mud plugs, diesel oil-bentonite plugs, diesel oil-cement plugs, bentonite-cement plugs, silica-clay plugs, time-setting clay plugs, and neat cement plugs.

A major obstacle to correcting lost circulation problems is locating the lost returns zone. Cementing at or near the casing seat eventually cures a high percentage of lost returns, particularly in shaly formations. Where the drilling mud is heavy and the crew has set a relatively deep protective string of casing, a squeeze job at the casing seat can reduce lost returns.

High Bottomhole Temperature

Temperatures in the well increase as the well gets deeper. This increase, or *geothermal gradient*, averages about 1°F per 60 feet (about 1°C per 33 metres), but may be considerably higher or lower. Figure 23 is a contour map of geothermal gradients in the southwestern U.S. In general, geologically older rocks of the interior United States and Canada are colder and have lower geothermal gradients than the younger Tertiary rocks of California and the Gulf Coast of Louisiana.

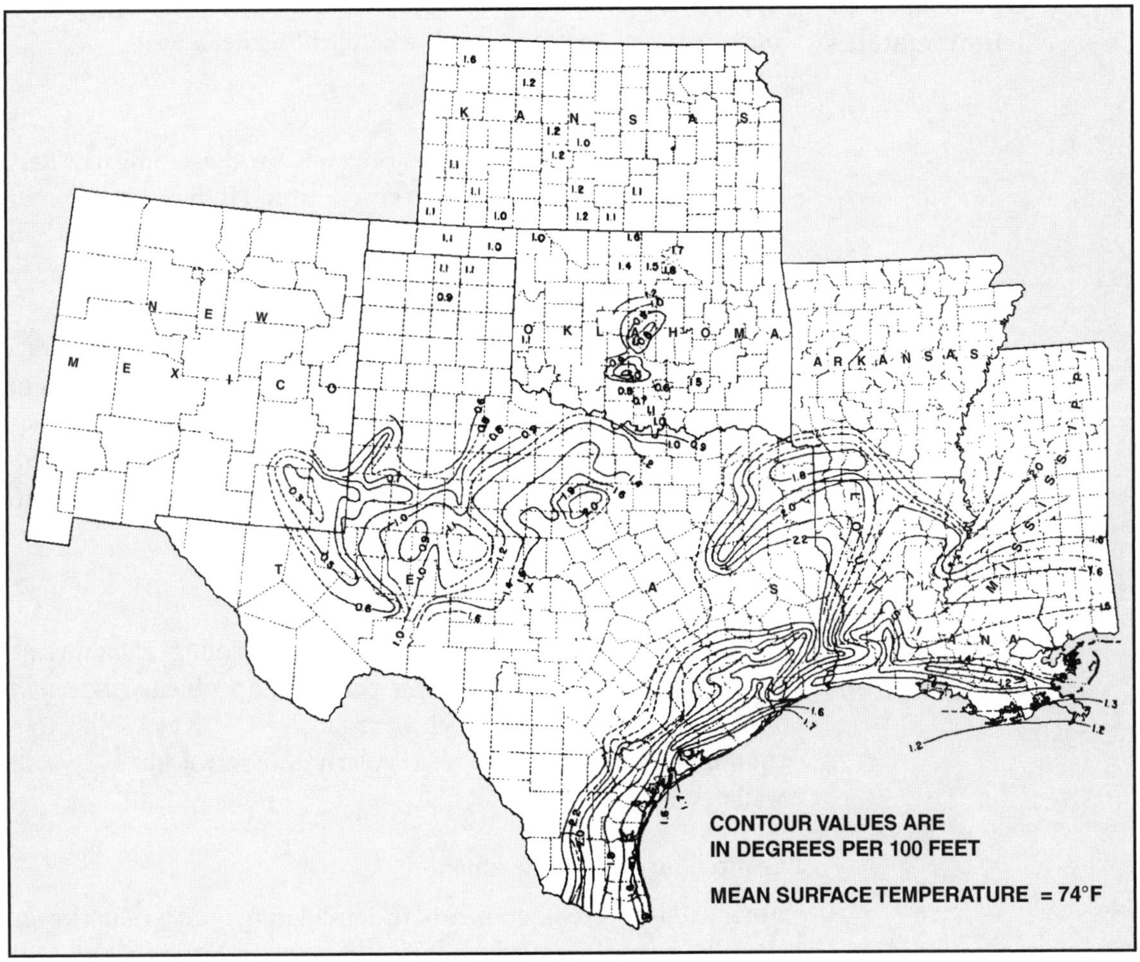

Figure 23. This contour map shows geothermal gradients in the southwestern U.S.

Beneficial Effects of Increased Temperatures

reactions, many components of drilling fluids that are relatively stable at surface temperatures react with one another at the elevated temperatures downhole. If the temperature is not too high, this can be beneficial. For example, downhole temperatures accelerate the reaction of chemicals added for flow control.

Detrimental Effects of Increased Temperatures

The effect of contaminants on the mud system becomes more severe as temperatures increase. Control of flow and fluid loss properties become more difficult in water-base muds, which is one reason operators may select oil muds when drilling deep wells.

Increased Viscosity

High bottomhole temperatures accelerate the thickening of water-base muds, which reduces penetration rates. High-viscosity muds can cause swabbing when tripping out and pressure surges when tripping in. In severe cases, repeated conditioning of the mud may be necessary before reaching bottom.

Mud thickening may prevent logging instruments and perforating guns from reaching bottom and cause drill stem test tools to become stuck. During workovers, expensive washover operations may be necessary if the mud used as a packer fluid has solidified. Excessive temperature can also cause fluid loss, which may lead to such problems as stuck pipe and shale trouble.

Solids Control

Solids control is also very important when drilling at high temperatures. High temperatures can increase the problems associated with high solids content, such as reduced ROP. Keep solids to a minimum, and add bentonite regularly to control filter loss and wall cake.

Controlling Other Properties

Fortunately, the temperature of the circulating mud is much lower than the temperature of the formation, and only part of the mud nears formation temperatures during trips. But high temperatures make the control of drilling fluid properties difficult because many dispersants and fluid loss additives break down and become ineffective as temperature increases (Table 10).

Table 10
Temperature Ranges of Common Drilling Muds

Mud Type	Temperature Range
Polyphosphate-treated muds	Up to 200°F (93°C)
Calcium treated muds	Up to 300°F (149°C)
Lignosulfonate muds	Up to 350°F (177°C)
Lignite muds	Over 350°F (177°C)
Oil muds	Up to 450°F (232°C)

For example, polyphosphate treated muds are only efficient at hole temperatures up to 200°F (93°C). Calcium muds, especially those having a high pH such as lime muds, can solidify at high temperatures. Low-pH calcium muds such as gyp-based muds are less inclined to solidify, but they are susceptible to fluid loss at temperatures around 300°F (149°C) because of the breakdown of fluid-loss additives such as starch and CMC.

Lignosulfonate muds are the most commonly used water-base muds in today's deep drilling, but they tend to break down at temperatures above 350°F (177°C). Above 350°F (177°C), lignite can be used in freshwater mud for thinning and filtration control. However, this type of mud is very susceptible to cement and saltwater contamination.

Oil muds are very popular in high-temperature wells. Unlike water-base muds, oil muds tend to thin as temperature increases. So controlling oil mud systems at high temperatures mainly involves keeping the viscosity and gel strength high enough to clean the hole and keep barite suspended.

Formation Pressure

Fluids contained in the pores of a permeable formation exert pressure, just as fluids in a borehole exert pressure. Many subsurface formations containing fluids outcrop on the surface (fig. 24). The formation is not likely to outcrop directly as shown in the figure, however. Instead, other permeable formations will likely overlie the formation and one of them will eventually outcrop. Nevertheless, the effect is the same: the outcropping rocks relieve pressure placed on the fluid-filled formation by the overburden—the overlying rocks. To understand the phenomenon, visualize a large, deep barrel or vat full of marbles. Now imagine pouring water into the marble-filled vat; the water easily flows in the spaces between the marbles and, if you add enough water, fills the spaces. The same thing happens in a porous and permeable formation. As long as a connection to the surface exists, the pressure the fluids exert is normal.

On the other hand, if the pressure created by the overburden is not relieved by a formation's being connected to the surface, then pressure in the formation can be abnormally high. Further, formation pressure can also be abnormally low, or subnormal. Subnormal formation pressures may be found in partially or totally depleted reservoir formations, formations at high elevations, and formations that outcrop downhill from a well.

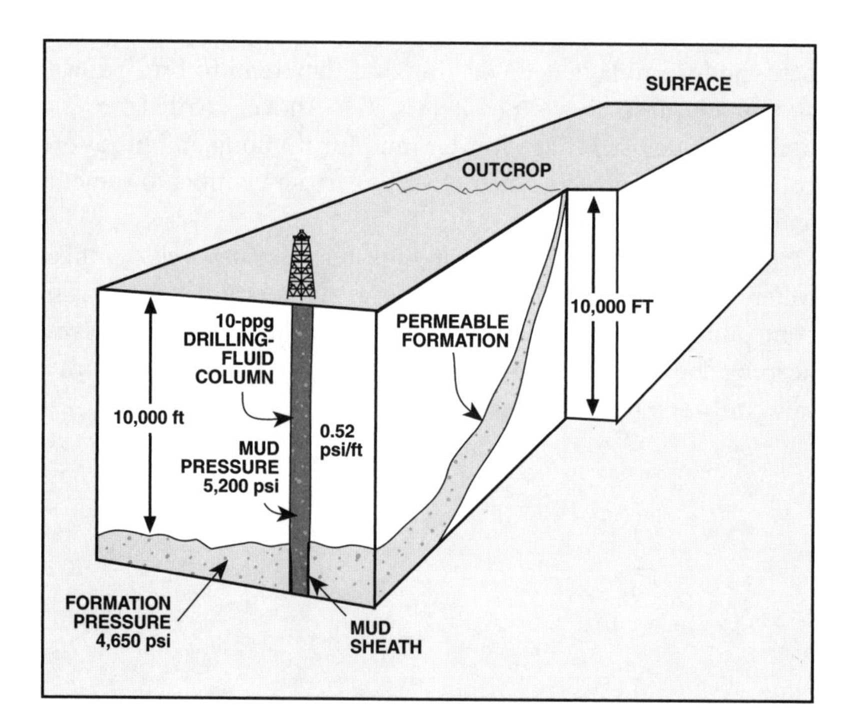

Figure 24. Many subsurface formations outcrop on the surface.

Whether normal or abnormal, fluids exert pressure with depth, and this pressure-with-depth relationship is usually measured in pounds per square inch per foot (psi/ft) or kilopascals per metre (kPa/m). The amount of pressure a fluid exerts with depth is the fluid's *pressure gradient*. Pressure gradient is a scale of pressure differences with a uniform variation in pressure from point to point. In other words, the deeper the fluids in a formation, the higher is the pressure they exert. Further, the denser a fluid is (the more it weighs), the more pressure it exerts at any given depth.

For example, consider a salt water that weighs 9 ppg (1,080 kg/m^3). This salt water has a pressure gradient of 0.468 psi/ft (10.584 kPa/m). Thus, if a hole 1,000 ft deep is filled with this salt water, the pressure it exerts at the bottom of the hole is 468 psi, because $0.468 \times 1{,}000 = 468$ psi. At the bottom of a 2,000-ft hole, this 9-ppg salt water exerts 936 psi. Similarly, if a hole 1,000 m deep is filled with this salt water, the pressure it exerts at the bottom of the hole is 10,584 kPa because $10.584 \times 1{,}000 = 10{,}584$ kPa. At the bottom of a 2,000-m hole, this 1,080- kg/m^3 salt water exerts 21,168 kPa.

Now suppose a hole 1,000 ft (1,000 m) deep is filled with a drilling mud with a density of 12.6 ppg (1,512 kg/m^3). This mud has a pressure gradient of 0.655 psi/ft (14.818 kPa/m), so at the bottom of a 1,000-ft hole, it exerts 655 psi, and at the bottom of a 1,000-m hole, it exerts 14,818 kPa. At the bottom of a 2,000-ft (2,000-m) hole, this 12.6-ppg (1,512-kg/m^3) mud exerts twice the pressure it does at 1,000 ft (1,000 m).

If you know the weight or density of the fluid, you can easily determine the pressure it exerts at any depth by multiplying its weight by the depth by a constant. The constant you use depends on how the mud weight is expressed. If mud weight is in ppg, the formula is—

$$PG_{psi/ft} = MW_{ppg} \times D_{ft} \times 0.052 \qquad \textit{Equation 1}$$

where

$PG_{psi/ft}$ = pressure gradient, psi/ft

MW_{ppg} = mud weight, ppg

D_{ft} = depth, ft

0.052 = a constant.

The constant 0.052 is derived from the fact that 1 cubic foot (ft^3) contains 7.48 U.S. gal. If a weightless cube measuring 1 ft on each side is filled with a substance weighing 1 ppg, the substance occupies 1 ft^3 or 7.48 gal. and weighs 7.48 lb, because $7.48 \text{ gal.} \times 1 \text{ ppg} = 7.48$ lb.

To find the pressure exerted on the bottom of the container, divide 7.48 lb by 144 in^2, because 144 in^2 are in 1 square ft (ft^2). Since 7.48 ÷ 144 = 0.05194, or 0.052, a column of fluid that is 1 ft high and weighs 1 ppg exerts 0.052 on bottom.

If the mud weight is in lb/ft^3, the formula is—

$$PG_{psi/ft} = MW_{lb/ft^3} D_{ft} \times 0.00694 \qquad \textit{Equation 2}$$

where

$PG_{psi/ft}$ = pressure gradient, psi/ft
MW_{lb/ft^3} = mud weight, lb/ft^3
D_{ft} = depth, ft
0.00694 = a constant.

The constant 0.00694 is also derived from the fact that a weightless container measuring 1 ft on each side contains 1 ft^3. If the container is filled with a substance that weighs 1 lb, then the substance weighs 1 lb/ft^3. To find the pressure in psi exerted on the bottom of the container, divide 1 lb by 144 in^2, because 144 in^2 are contained in 1 ft^2. Since 1 divided by 144 equals 0.0069444, or 0.00694, a column of fluid that is 1 ft high and weighs 1 lb/ft^3 exerts 0.00694 psi on bottom.

If the mud weight is in kg/m^3, the formula is—

$$PG_{kPa/m} = MW_{kg/m^3} D_m \times 0.0098 \qquad \textit{Equation 3}$$

where

$PG_{kPa/m}$ = pressure gradient, kPa/m
MW_{kg/m^3} = mud weight, kg/m^3
D_m = depth, m
0.0098 = a constant.

The constant 0.0098 is derived from the fact that 1 pascal (Pa) is equal to 1 newton (N) of force applied to 1 square metre (m^2) and that a weightless cube measuring 1 m on each side contains 1 cubic metre (m^3). N is a measure of the gravitational force exerted by a mass. One N exerts 9.8 kilograms (kg) of force on the bottom of 1 m^3. Because the measurement is in kPa, then one must divide 9.8 by 1,000 because there are 1,000 Pa in 1 kPa. Thus, a column of fluid that is 1 m high and has a force of 1 N on the bottom exerts 0.0098 kPa.

As mentioned earlier, oil people classify the pressures formation fluids exert as normal, subnormal, and abnormally high. Most consider a pressure gradient of 0.433 to 0.468 psi/ft (9.786 to 10.577 kPa/m) as normal. That is, if the formation fluids do not exert more or less than 0.433 to 0.468 psi/ft (9.786 to 10.577 kPa/m) at any given depth, then formation pressure is considered to be normal.

Thus, a formation containing salt water that exerts 0.465 psi/ft (10.509 kPa/m) is normally pressured. This normally pressured formation exerts 46.5 psi at 100 ft, 465 psi at 1,000 ft, and 4,650 psi at 10,000 ft. Similarly, a normally pressured formation salt water that exerts 10.509 kPa/m, exerts 105.09 kPa at 10 m, 1,050.9 kPa at 100 m, and 10,509 kPa at 1,000 m.

Subnormal pressure zones can create lost circulation problems. Kicks are a hazard in normal and abnormally high pressure zones, particularly the latter. Theoretically, abnormally high pressures range from 0.468 to 1.00 psi/ft (10.587 to 22.6 kPa/m). Many operators, however, consider abnormal formation pressures to be those over 0.620 psi per foot (14.012 kPa/m), or a pressure that a mud weight of 12.0 ppg (1,440 kg/m^3) will overcome.

Circulating Pressure Loss

The ability of a drilling fluid to control downhole formation pressure depends primarily on hydrostatic pressure. The greater the density of the mud and the deeper the column, the greater is the hydrostatic pressure. However, circulating pressure—the pressure put out by the mud pump—also affects formation pressure control because of annular pressure loss. As the mud moves out of the pump, up the standpipe, down the drill string, out of the bit, and up the annulus, it loses circulating pressure. It loses circulating pressure because of friction. The friction is caused by the mud's moving against the pipe walls and by internal friction—particles in the mud moving against each other. Friction reduces circulating pressure. This phenomenon is called friction loss, pressure loss, or circulating pressure loss. Normally, the greatest loss of circulating pressure occurs as the mud jets out of the bit nozzles. The circulating pressure left to move mud up the hole after the mud leaves the bit is relatively small. And, by the time the mud reaches the surface, all of the circulating pressure is used up. Some factors that govern how much circulating pressure is lost because of friction are the flow rate of the mud; the viscosity of the mud; the size and length of the rig's surface equipment; the configuration, length, and size of the drill string; the size of the bit nozzles; and the size of the hole. In general, the smaller the inside diameter of the pipe through which the mud is flowing, the greater are the friction losses. Also, the faster the mud is flowing, the greater are the friction losses.

As an example of circulating pressure losses using English units, suppose that a pump puts out 2,400 psi at the surface. As the mud travels up the standpipe and through the swivel of a particular rig, suppose that it loses 50 psi. The mud then travels down the drill string to the bit where it loses an additional 550 psi. The mud then jets out of the bit nozzles where it loses 1,700 psi. At this point, the mud has only 100 psi of pressure left to move up the annulus to the surface, because 50 + 550 + 1,700 = 2,300, and 2,400 – 2,300 = 100 psi. At the surface, circulating pressure is zero because all the circulating pressure has been lost, or used up, as the mud moves up the annulus (fig. 25).

Using SI units, suppose the pump puts out 16,500 kPa and that the mud loses 350 kPa in the standpipe and swivel, 3,500 kPa down the drill string, and 12,000 kPa out of the bit nozzles. The mud now has only 650 kPa of pressure left to move up the annulus, because 350 + 3,500 + 12,000 kPa = 15,850, and 16,500 – 15,850 = 650 kPa.

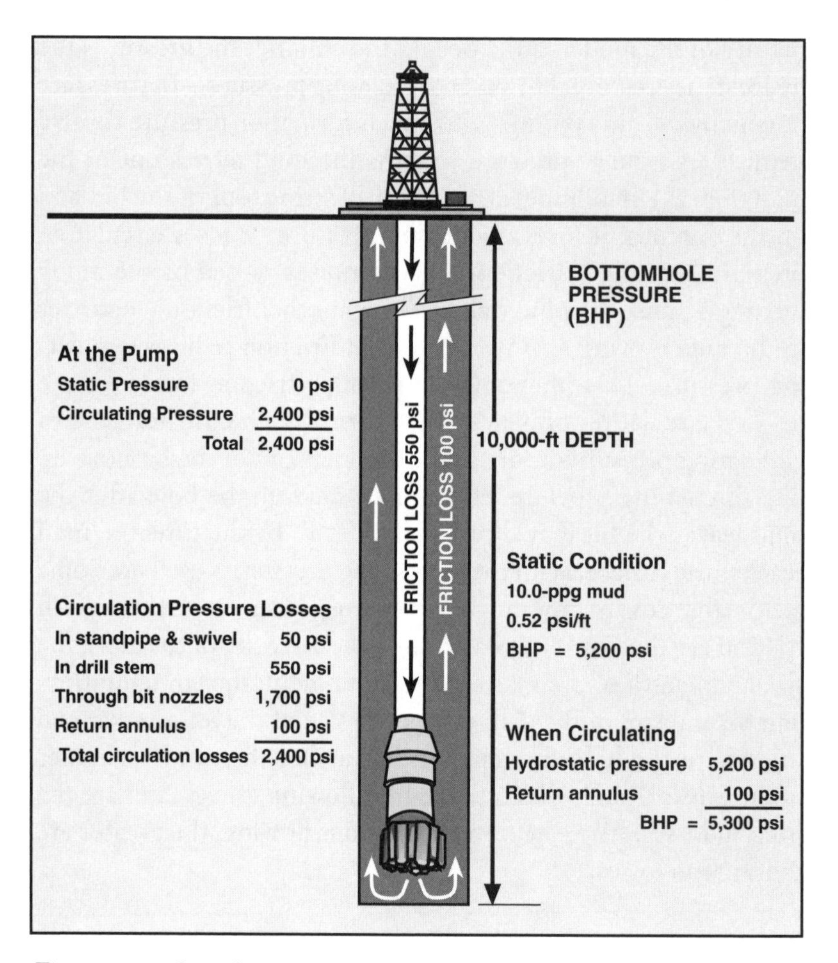

Figure 25. Circulating pressure losses

At the surface, circulating pressure is zero because all the circulating pressure has been lost, or used up, as the mud moves up the annulus.

A critical thing to remember about pressure loss in the annulus is that you must add it to the hydrostatic pressure at the bottom of the hole when mud is circulating to determine how much pressure is actually being exerted at the bottom of the hole. For example, if the hydrostatic pressure the mud exerts on bottom is 3,763 psi (25,946 kPa) and the pressure loss of the mud in the annulus is 50 psi (345 kPa), then bottomhole pressure while circulating is 3,763 + 50 = 3,813 psi (25,946 + 345 = 26,291 kPa). While 50 psi (345 kPa) is not much pressure when compared to the thousands of psi (kPa) of hydrostatic pressure, this small value can be crucial. As an example, suppose that a formation exerts 3,800 psi (26,220 kPa). While circulating, bottomhole pressure is 3,813 psi (26,310 kPa), which includes 50 psi (345 kPa) of annular pressure loss. To make a connection, the driller shuts down the mud pump. Shutting down the pump gets rid of the 50 psi (345 kPa) of annular pressure loss. Consequently, bottomhole pressure is now 3,763 psi (25,965 kPa), which is 37 psi (255 kPa) less than formation pressure. The formation can now kick—that is, fluids in the formation can enter the wellbore because bottomhole hydrostatic pressure is less than formation pressure.

Swabbing

When the crew pulls pipe up the wellbore, the hydrostatic pressure below the bit drops, which can allow formation fluids to enter the well. The bit and pipe's pulling formation fluids into the well is *swabbing*. The faster the driller raises the pipe, the worse swabbing becomes. Swabbing explains why mud is sometimes cut with salt water or gas after a trip. How bad swabbing is depends on the depth of the pipe, the gel strength of the mud, and the speed at which pipe is pulled from the hole. A balled bit and large-diameter drill collars can also worsen swabbing.

If the mud weight creates a hydrostatic pressure that is only slightly higher than bottomhole formation pressure, and there is no danger of lost circulation, some operators prefer to increase the mud weight a small amount to help combat swabbing. Such a mud weight increase is often called a *trip margin*. One way to calculate how much weight to add is to take the actual mud weight required to balance formation pressure and increase this value by a safety factor. Some operators use a value of 0.3 ppg (36 kg/m^3); others weight up enough to increase bottomhole hydrostatic pressure by 200 to 400 psi (1,380 to 2,760 kPa).

Well Kicks

If formation pressure is greater than hydrostatic pressure, formation fluid enters the well, causing a kick. A kick must be detected and correctly controlled to avoid a blowout. All crew members therefore need to recognize signs of a kick. While many kick signs can occur, three clear-cut ones while drilling are (1) the volume of fluid returning from the hole increases and no one has changed pump speed; (2) the level of mud in the mud tanks increases and no one has added mud to the system; and (3) the pump is shut down, no pill of heavy mud or other heavy fluid is in the drill string, and, after a short wait, mud continues to flow from the well. Most rigs come equipped with sensitive instruments to accurately measure mud return flow and tank volume. When return flow exceeds a normal set point, and when mud-tank volume increases above a normal set level, alarms alert the driller to a possible kick. Note, too, that the third sign is also called a flow check and can be used to verify that a kick has occurred if doubt exists.

To prevent kicks, keep the hole full of mud of the proper weight at all times and use mud whose weight is heavy enough to develop enough hydrostatic pressure to overcome formation pressure. Properly killing a kick involves closing the blowout preventers (fig. 26), circulating the kick out of the well through a choke, and circulating mud into the well with a weight sufficient to overcome formation pressure.

Kicks while Drilling

Kicks that occur when the bit is on bottom and drilling, and the pump is circulating drilling mud, are often easy to recognize. When a kick occurs, the volume of mud in the tanks and drilling rate usually increase. (An increase in the drilling rate is called a *drilling break*.) The formation fluids enter the hole to join the mud already there, which shows up as an increase in return flow and a tank gain on the surface. Further, the kick fluids reduce bottom-hole pressure, which makes it easier for the cuttings to move off bottom. With the cuttings removed, the bit cutters are exposed to more uncut formation, which increases ROP. At the same time, the pump pressure may decrease and pump speed increase. The kick fluids in the annulus lighten the annular mud column, making it easier for the pump to move the mud.

Figure 26. Blowout preventer stack on a land rig

Well Kicks When Pump Is Shut Down

Sometimes, no signs of a kick occur during circulation, but, when the driller shuts down the pump and waits long enough for the pump pressure to dissipate throughout the circulating system, mud continues to flow out of the return line. If a well flows with the pumps off but no signs of a kick occurred while circulating, it is possible that hydrostatic pressure alone is not enough to overcome formation pressure. Formation fluid may not enter the wellbore during circulation because hydrostatic pressure plus the pressure added to the bottom of the hole by annular pressure loss is enough to control formation pressure.

If the drill string is not slugged with a heavy fluid, which means that the heavy fluid could U-tube out of the drill string and force mud up the annulus, and if the well flows with the pumps off, it usually shows that kick fluids are entering the well. If the well flows with the pumps off, the well should be shut in. With the well shut in, and the pumps off, the driller can firmly establish that a kick has occurred by checking the standpipe pressure gauge or the drill pipe pressure gauge. If either gauge indicates pressure, with the pumps off and the well completely closed in, then the well has kicked and steps must be initiated to kill the well.

To summarize the steps drillers can take whenever there is doubt about whether a well has kicked include: (1) pick up the bit off bottom; (2) shut down the mud pump; (3) wait enough time to ensure that the mud is no longer flowing in reaction to being pumped; (4) make sure the well is completely open so that no residual pump pressure is trapped in the system; (5) shut the well in completely; (6) check to see if the drill pipe or standpipe pressure gauge indicates pressure. If the drill pipe gauge indicates pressure, then the well has kicked.

Kicks When Coming Out of the Hole

One cause of kicks when tripping pipe out of the hole is swabbing. What frequently happens is that each time the crew pulls a stand of pipe a little formation fluid is swabbed in. Soon, after several stands have been pulled, enough formation fluids have entered the hole to cause a kick. (Remember that a kick is the entry of enough formation fluids into the wellbore so that when the well is completely shut in, pressure appears on the drill pipe pressure gauge.)

When pulling pipe out of the hole, the crew must add mud to the hole to replace the volume occupied by the pipe. Otherwise, the mud level in the hole falls to the point that it no longer exerts enough bottomhole pressure to overcome formation pressure. Therefore, it is important to keep the hole full of mud. Keeping the hole full, however, is only part of the story. It is one thing to know the hole is full, but what is it full of—mud alone or mud and formation fluids? Calculating the volume of mud pumped into the hole and making sure that this amount corresponds to the pipe displacement is therefore critical. If the hole fails to take enough mud—that is, if mud flows out the return line before the calculated amount of mud to replace drill string volume is pumped in—then it is probable that formation fluids have been swabbed into the hole.

Because the crew must keep accurate track of the amount of mud they put into the hole to replace the drill string, many rigs have a trip tank (fig. 27) with an accurate, finely graduated gauge that shows how much mud has been pumped in to replace the string. Trip tanks are often graduated in ¼-bbl, gallon-, decalitre-, or litre-increments to ensure that the driller can keep very accurate track of the volume of mud pumped in to replace drill pipe and drill collar stands. When a trip tank is combined with a trip sheet (fig. 28), which shows the displacement volume of each drill string element, rig crews should have no difficulty keeping track of exactly how much replacement mud has gone into the hole.

Figure 27. Trip tank

RIG: ____________________ DATE: ____________________

WELL: ____________________ TIME: ____________________

DRILLER: ________________ **TRIP SHEET** DEPTH: ____________________

REASON FOR THE TRIP: ____

Number of stands to have top of DC's one DP stand below BOP's: ________________

PULL ON:	✓
EVEN	
SINGLE	
DOUBLE	

DISPLACEMENT:	DC1	DC2	OTHER	HWDP	DP1	DP2
Size						
bbl/ft or						
bbl/stand						
x ft or stands						
= Vol. (bbls)						

STAND NO.	TRIP TANK GAUGE	CALCULATED Hole Fill (bbls) per Increment	MEASURED Hole Fill (bbls)		DISCREPANCY		REMARKS
			per Increm'	Accumul.	per Increm'	Accumul.	
0							

Figure 28. Trip sheet

Kicks Going into the Hole

The opposite of swabbing is surging. Surging occurs when the driller lowers the drill string into the hole. Pressure surges can be minimized by lowering the drill string slowly. One danger with surges is that they can be strong enough to fracture a formation in the open part of the hole (the part without casing). If the fracture is severe enough, lost circulation can occur. As discussed previously, lost circulation can cause the height of the fluid column to fall, thus decreasing hydrostatic pressure and allowing formation fluids to enter the wellbore. When running pipe into the hole, the driller can recognize that lost circulation has occurred when the drill string fails to displace from the hole an amount of mud equal to the volume of the drill string put into the hole.

Shutting In a Well

When a well kicks, the first step in regaining control of the well is to shut it in. Operators and contractors usually have their own specific steps to be followed when shutting in a well. The following steps are offered merely as a guide; crew members should check with their supervisor to ensure that they are familiar with the procedures on their particular rig.

1. Sound an alarm to alert the crew and supervisory personnel. Raise the drill string to get the bit off bottom and minimize the possibility of stuck pipe.
2. Stop the pump.
3. Wait a short time for pump pressure to dissipate throughout the mud system. Then, shut in the well completely by closing a blowout preventer and the choke.
4. Check to see how much, if any, pressure is indicated by the drill pipe (standpipe) pressure gauge. If the drill string has a float (a valve that prevents mud from flowing back up the drill string) in it, the drill pipe pressure gauge cannot indicate pressure. In this case, the driller should check the casing pressure gauge. If it indicates pressure, a kick has occurred. Do not, however, use shut-in casing pressure to determine how much to weight up the mud. Fluids in the casing annulus are a mixture of mud and kick fluids of unknown volume, composition, and density. Consequently, shut-in casing pressure (SICP) cannot be used to determine further calculations, such as mud-weight increase.

Shut-In Drill Pipe Pressure (SIDPP)

Shut-in drill pipe pressure (SIDPP) is essential in calculating the mud-weight increase needed to kill the well. A barrier to obtaining SIDPP directly is that many operators require that a float valve be run in the drill string to prevent kick fluids from entering the string when the mud pump is shut down, as it is during trips in or out of the hole. A float valve is a check valve: it allows flow in one direction only. Crew members usually install the float near the bottom of the string just above the bit. Pump pressure from above opens the float, but pressure from below (where kick pressure comes from) closes the float and prevents upward flow. With the float closed against pressure from below, no pressure indication shows up on the drill pipe pressure gauge. Because the float prevents reading pressure on the drill pipe gauge, it is necessary to determine SIDPP indirectly. One way to determine SIDPP is for the driller to start the pump slowly and carefully watch the drill pipe pressure gauge. At some point, pump pressure opens the float and drill pipe pressure drops. The pressure at the point where the float opens is SIDPP.

Determining Mud-Weight Increase

Whether SIDPP is obtained directly or indirectly, it is used to determine how much to increase the mud weight to balance formation pressure. One formula used in the English system of measurement is—

$$W_r = W_i + \frac{SIDPP_{psi}}{0.052\ TVD_{ft}} \qquad \textit{Equation 4}$$

where

W_r = required mud-weight increase, ppg
W_i = initial mud weight, ppg
$SIDPP_{psi}$ = shut-in drill pipe pressure, psi
TVD_{ft} = true vertical depth of hole or length of pipe, ft.

It is important to note that the true vertical depth (TVD) of the hole or true vertical length of the drill string must be used to obtain the correct mud-weight increase needed to balance the formation. In directional and horizontal wells, the measured depth (MD) of the hole (the depth obtained by measuring along the well's actual path) may be quite different from TVD. For instance, suppose a well is drilled vertically to a depth of 3,000 ft (914.4 m). At this depth, the crew kicked it off vertical and drilled it on a slant to a TVD of 6,000 ft (1,828.8 m). While TVD is 6,000 ft (1,828.8 m), MD in this example is 8,000 ft (2,438.4 m) (fig. 29).

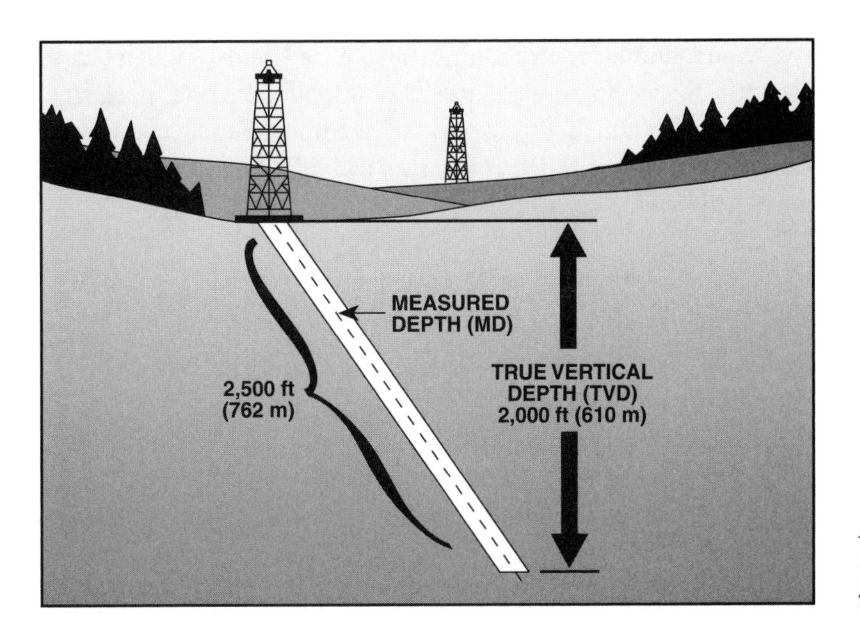

Figure 29. Measured depth (MD) is different from true vertical depth (TVD).

As an example in determining the required mud-weight increase using the English system, suppose a well being drilled with a 12.2-ppg mud has a TVD of 8,976 ft and kicks. After the crew completely shuts the well in, SIDPP reads 350 psi. What is the required mud-weight increase to balance formation pressure?

$$W_r = 12.2 \text{ ppg} + \frac{350 \text{ psi}}{0.052 \times 8{,}976 \text{ ft}}$$

$$= 12.2 + \frac{350}{466.75}$$

$$= 12.2 + 0.750$$

$$W_r = 12.95 \text{ or } 13.0 \text{ ppg.}$$

In this example, crew members add enough weighting material to the mud to increase its weight from 12.2 to 13.0 ppg.

In the SI system of measurement, one formula used to determine the required mud-weight increase is—

$$W_r = W_i + \frac{SIDPP_{kPa}}{0.0098\ TVD_m} \qquad \text{Equation 5}$$

where

W_r = required mud-weight increase, kg/m^3

W_i = initial mud weight, kg/m^3

$SIDPP_{kPa}$ = shut-in drill pipe pressure, kPa

TVD_m = true vertical depth of hole, m.

As an example in determining the required mud-weight increase using the SI system, suppose a well being drilled with a 1,464 kg/m³ mud has a TVD of 2,736 m and kicks. After the crew completely shuts the well in, SIDPP reads 2,413 kPa. What is the required mud-weight increase to balance formation pressure?

$$W_r = 1{,}464 \text{ kg/m}^3 + \frac{2{,}413 \text{ kPa}}{0.0098 \times 2{,}736}$$

$$= 1{,}464 + \frac{2{,}413}{26.81}$$

$$= 1{,}464 + 90.0$$

$$W_r = 1{,}554 \text{ kg/m}^3$$

In this example, crew members add enough weighting material to increase the mud weight from 1,464 to 1,554 kg/m³.

Remember that equations 4 and 5 result in a mud weight that balances formation pressure. It does not overbalance it. Some operators prefer to add a few tenths of a pound per gallon (a few kg/m³) as a safety factor. For example, if the new mud weight required to balance the formation is 13.0 ppg (1,560 kg/m³), an operator may order the mud to be weighted to 13.3 ppg (1,596 kg/m³). On the other hand, many operators prefer not to add any additional weight to overbalance formation pressure. Reasons include—

1. When the new, heavier mud is circulated into the well and up the annulus, annular pressure loss plus hydrostatic pressure provide an overbalance.
2. Using the lightest possible mud weight results in the best ROP.
3. Some formations fracture if the mud weight is increased more than required. Fractured formations lead to lost circulation with its problems.

Killing a Well

Shutting in the well and determining the mud weight required to control the kick are initial steps in controlling a kick. Once shut in, however, the crew must take steps to circulate the intruded kick fluids out of the well and to circulate the new, heavier mud into the well and back to the surface. Many methods are available for performing these operations, but one popular one is the driller's method.

Driller's Method

In the driller's method, the well is shut in, SIDPP and SICP are noted and recorded, and the kick is circulated out of the well while maintaining a constant bottomhole pressure with the choke. By keeping the pump at a constant speed and not allowing it to vary, and by adjusting the size of the choke opening, bottomhole pressure can be maintained at the correct value to prevent more kick fluids from entering the hole. At the same time, bottomhole pressure is not kept so high as to risk fracturing a formation.

One of the keys to successful well control is to determine a kill-rate pump speed and pressure before the well kicks. The kill-rate speed is usually well below the speed normally used when drilling. Indeed, many operators require that the driller record two or three kill rates. Often, drillers may begin with one kill rate at about one-half the normal rate and then determine two others at one-third and one-fourth the normal rate. For example, suppose the normal pump rate and pressure is 60 strokes per minute (spm) at 2,200 psi (15,180 kPa). The driller then reduces the pump rate to 30 spm and records pump pressure at that speed; as an example, let's say it is 550 psi (3,795 kPa). Next, the driller reduces pump speed to 20 spm and records the pressure at that speed; let's say it is 250 psi (1,725 kPa). Finally, the driller reduces pump speed to 15 spm and records the pressure at that speed; in this case, 150 psi (1,035 kPa). These reduced pressures are called kill rate pressures (KRPs).

Reducing the pump rate and establishing KRP to circulate a kick out of the well and pump the new, heavier mud in is necessary for several reasons. First, reducing the pump speed reduces circulating friction losses so that circulating pressures are less likely to cause excessive pressure on exposed formations. Second, it gives the rig crew more time to mix the heavier mud needed for the kill. Third, a slow rate reduces wear and tear on the pump; a pump running at slow speed has an easier time of it than a pump running fast. Fourth, it gives those involved in the killing procedure more time to react to problems. And fifth, it allows adjustable chokes to work within their orifice ranges.

The following is one step-by-step procedure for the driller's method of well control. Be aware that operators and contractors may require steps that vary from the ones given here. Crew members should be familiar with and adhere to the steps required on their particular rig.

1. When the kick occurs, shut in the well completely in compliance with the operator's and contractor's procedures.
2. Record SIDPP, SICP, and mud tank (pit) gain on a well-control work sheet (fig. 30).

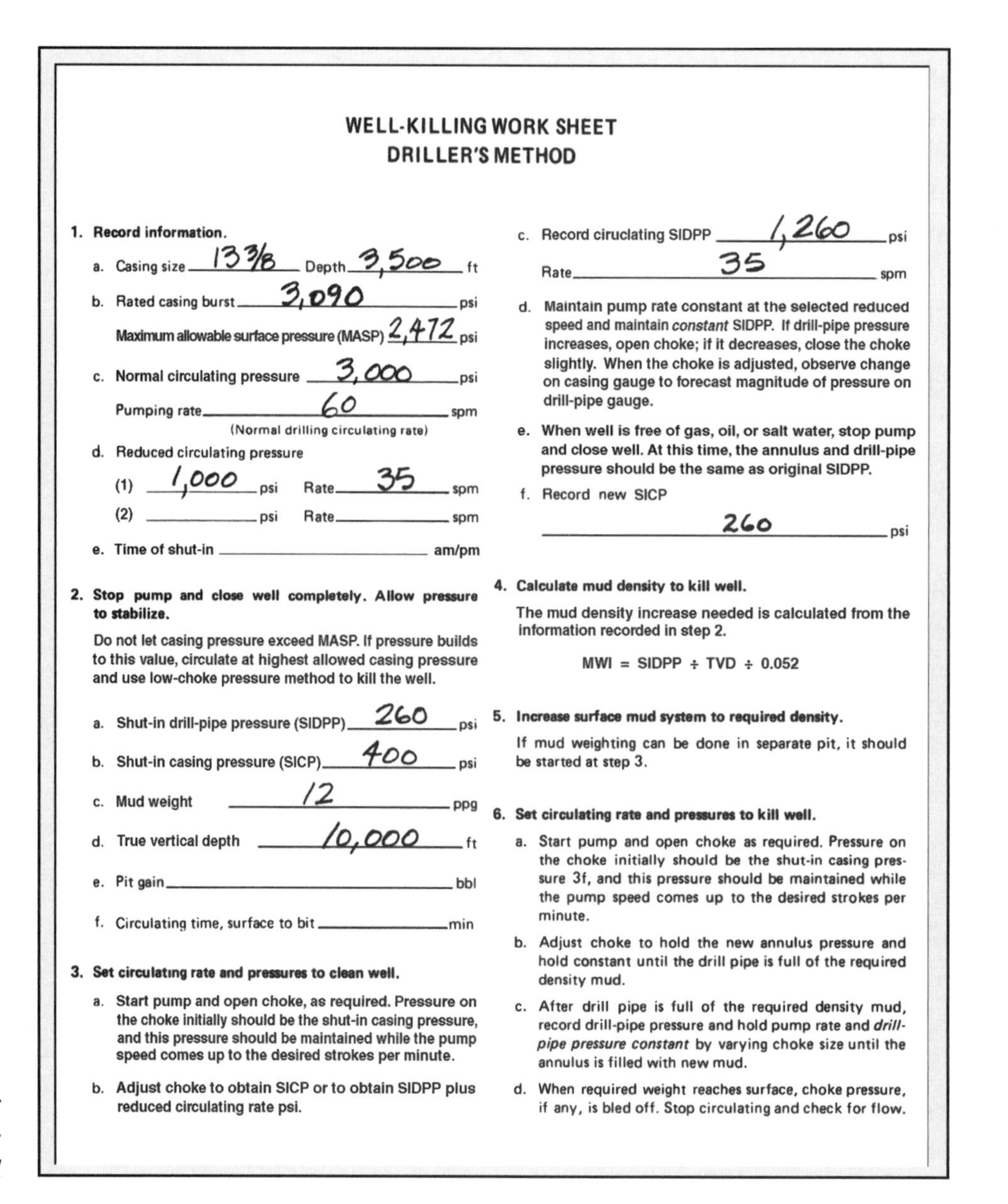

WELL-KILLING WORK SHEET
DRILLER'S METHOD

1. Record information.
   a. Casing size 13 3/8 Depth 3,500 ft
   b. Rated casing burst 3,090 psi
      Maximum allowable surface pressure (MASP) 2,472 psi
   c. Normal circulating pressure 3,000 psi
      Pumping rate 60 spm
      (Normal drilling circulating rate)
   d. Reduced circulating pressure
      (1) 1,000 psi Rate 35 spm
      (2) ____ psi Rate ____ spm
   e. Time of shut-in ____ am/pm

2. Stop pump and close well completely. Allow pressure to stabilize.
   Do not let casing pressure exceed MASP. If pressure builds to this value, circulate at highest allowed casing pressure and use low-choke pressure method to kill the well.
   a. Shut-in drill-pipe pressure (SIDPP) 260 psi
   b. Shut-in casing pressure (SICP) 400 psi
   c. Mud weight 12 ppg
   d. True vertical depth 10,000 ft
   e. Pit gain ____ bbl
   f. Circulating time, surface to bit ____ min

3. Set circulating rate and pressures to clean well.
   a. Start pump and open choke, as required. Pressure on the choke initially should be the shut-in casing pressure, and this pressure should be maintained while the pump speed comes up to the desired strokes per minute.
   b. Adjust choke to obtain SICP or to obtain SIDPP plus reduced circulating rate psi.
   c. Record circulating SIDPP 1,260 psi
      Rate 35 spm
   d. Maintain pump rate constant at the selected reduced speed and maintain *constant* SIDPP. If drill-pipe pressure increases, open choke; if it decreases, close the choke slightly. When the choke is adjusted, observe change on casing gauge to forecast magnitude of pressure on drill-pipe gauge.
   e. When well is free of gas, oil, or salt water, stop pump and close well. At this time, the annulus and drill-pipe pressure should be the same as original SIDPP.
   f. Record new SICP 260 psi

4. Calculate mud density to kill well.
   The mud density increase needed is calculated from the information recorded in step 2.

   MWI = SIDPP ÷ TVD ÷ 0.052

5. Increase surface mud system to required density.
   If mud weighting can be done in separate pit, it should be started at step 3.

6. Set circulating rate and pressures to kill well.
   a. Start pump and open choke as required. Pressure on the choke initially should be the shut-in casing pressure 3f, and this pressure should be maintained while the pump speed comes up to the desired strokes per minute.
   b. Adjust choke to hold the new annulus pressure and hold constant until the drill pipe is full of the required density mud.
   c. After drill pipe is full of the required density mud, record drill-pipe pressure and hold pump rate and *drill-pipe pressure constant* by varying choke size until the annulus is filled with new mud.
   d. When required weight reaches surface, choke pressure, if any, is bled off. Stop circulating and check for flow.

Figure 30. Worksheet for the driller's method

3. To start circulating, open the choke, slowly bring the pump up to kill rate, and hold SICP at a constant value, which is usually the original shut-in reading, by adjusting the choke. Open the choke to reduce SICP and close the choke to increase SICP.
4. When the pump is at kill-rate speed, observe the drill pipe pressure gauge; it shows initial circulating pressure (ICP). ICP is equal to KRP plus SIDPP. Hold ICP constant by opening or closing the choke. Do not allow the pump to change speed.
5. Circulate the kick out, holding SIDPP constant at ICP. Keep in mind that a time lag exists when adjusting the choke. A choke adjustment does not immediately show up on the drill pipe pressure gauge. The length of the time lag depends on the length (depth) of the drill string. The longer the drill string is, the longer it takes for the adjustment to appear on the gauge. A rule of thumb is to expect about 1 second of delay per 1,000 ft (300 m) of depth. For example, in a 10,000-ft (3,000-m) well, the choke operator would expect a 20-second delay before a choke adjustment appears on the drill pipe gauge (10 seconds down the pipe and 10 seconds up the annulus). SICP response time, on the other hand, is quite rapid because annular pressure changes occur at the surface where casing pressure is read. The rapid casing pressure response can be used to adjust SIDPP. For example, say it is necessary to decrease SIDPP by 100 psi (690 kPa). The choke operator can observe the casing pressure gauge while opening the choke until SICP drops by 100 psi (690 kPa). In the example 10,000-ft (3,000-m) well, SIDPP should also drop by 100 psi (690 kPa) about 20 seconds later. When all of the kick influx has been pumped out and the well shut in, SIDPP and SICP should be equal to the SIDPP noted when the kick occurred.
7. Stop the pump, completely close the choke, and finish mixing kill-weight mud, if necessary.
8. When the mud tanks (pits) are full of kill-weight mud ready to be pumped into the well, open the choke and slowly bring the pump up to the kill rate; hold SICP constant at the shut-in reading until the new mud gets to the bit. As new mud fills the drill string, ICP will slowly decrease toward final circulating pressure (FCP).

9. When kill-weight mud reaches the bit, stop observing SICP and begin observing SIDPP. Hold SIDPP constant by choke adjustments at the new indicated value—FCP—until kill-weight mud appears at the mud return line or choke. Keep the pump speed constant.
10. Stop the pumps and shut in the well completely to ensure that the well is dead. If dead, both SIDPP and SICP will read 0. As a final check, open the well completely and check to see if the well flows with the pumps off. If the well flows, shut it in, read and record SIDPP and SICP, and begin the driller's method again.

Corrosion

The drill string or casing can corrode during drilling operations, particularly when the drilling fluid is a clear brackish or salt fluid. The most common corrosives are oxygen, carbon dioxide, and hydrogen sulfide.

Oxygen

Oxygen is a prime corroder of metals. Heat, water, and salt, all found downhole, increase oxygen's ability to destroy equipment. As the drilling fluid passes through the shaker screens, agitators, and hopper discharge, it absorbs oxygen from the air. Clear, low-solids fluids absorb air quite readily. Viscous, high-gel muds absorb it less readily, but, once absorbed, the air is more difficult to remove. Unless some means such as a degasser removes the air with its oxygen, this type of corrosion is difficult to combat. Methods that have some success include—

- using pipe manufactured with a plastic surface coating on the inside;
- coating pipe inside and out with a liquid corrosion inhibitor;
- treating the mud with lignosulfonate or quebracho;
- maintaining the mud at an alkaline pH, above 9.5;
- using a degasser; and
- adding a chemical oxygen scavenger such as sodium sulfite (this may be too costly since the mud picks up air continuously as it circulates, and the crew must add the chemical continuously too).

Carbon Dioxide

Some formation gases contain large amounts of carbon dioxide, or CO_2. Carbon dioxide by itself does not corrode metal, but its presence increases downhole corrosion in two ways. First, it reacts with water in the mud to form carbonic acid. Carbonic acid in turn reacts with the alkaline mud, reducing or completely eliminating its alkalinity. If the pH drops too low (that is, the mud becomes more acidic), the carbonic acid can also react with the iron in the steel drill string. So the combination of low pH and the presence of carbonic acid, both brought about by CO_2, corrodes the metal. One treatment for CO_2 is to add lime (calcium hydroxide) to react with any acid formed, and caustic soda, to maintain an alkaline pH and good flow properties.

Sulfides

One of the most dangerous conditions in drilling is the presence of hydrogen sulfide (H_2S) in a gas kick. Hydrogen sulfide is a threat both to equipment downhole and to people. As for its threat to personnel, table 11 shows it effects at concentrations from 100 parts per million (ppm) to 1,000 ppm. It is a very toxic gas and can be a serious threat to life if proper steps are not taken to deal with it.

H_2S is explosive in the presence of air. Downhole, hydrogen sulfide causes *sour corrosion*, a very severe type that makes metal and rubber so brittle that they crack. High-strength steels are particularly susceptible to H_2S corrosion.

Table 11
Effects of Hydrogen Sulfide on Humans

Concentration	Reaction
100 ppm	Coughing, eye irritation, loss of smell after 2–5 min
200 ppm	Marked eye and respiratory tract irritation after 1-hr exposure
500 ppm	Loss of consciousness and possibly death in 30 min to 1 hr
700 ppm	Rapid unconsciousness, cessation of breathing, and death
1,000 ppm	Unconsciousness with early cessation of breathing and death in a few minutes even if removed to fresh air at once.

Sulfides can also exist in an acidic mud as sodium sulfide or sodium bisulfide, both dangerous substances. These sulfides are solids that dissolve in water and release H_2S when they contact an acid.

Preventing H_2S Corrosion

As soon as sulfide enters a mud stream, the corrosion process begins, so the crew should begin corrosion-prevention measures without delay. The crew may need to add large amounts of caustic soda (2.35 parts of caustic soda to 1 part of sulfide) to bring the mud's pH to 9.5 or higher, which decreases but does not eliminate corrosion. The greatest damage occurs when sulfide is present as H_2S.

A degasser can pull H_2S out. This is particularly necessary when using clear water or a brine drilling fluid. The operator must arrange to safely burn off any H_2S gas removed by a degasser to protect the crew. Copper carbonate added to a mud precipitates H_2S as harmless copper sulfide.

To protect drill pipe, the crew can use a filming amine corrosion inhibitor. Apply the inhibitor directly to the inside and outside of the pipe as though it were a paint. A good inhibitor wets the steel and adheres strongly to it. It also remains as a film on the pipe for a long time. Amine inhibitors tend to adhere to clay particles in the mud as well as the steel, so mud composition may affect the results. Use pipe with a plastic-coated interior.

Safety Precautions

Most mud additives are relatively harmless, but a few can cause burns or toxic reactions and require special precautions. The degree of danger involved with handling these materials varies. The ones listed below are especially dangerous, but many others can be mildly toxic or irritating. Handle all chemicals with care, and avoid breathing the dust from dry materials.

Caustic Soda

Caustic soda, or sodium hydroxide (NaOH), commonly known as lye, is very strongly alkaline and can cause serious burns to the skin. It also makes holes in leather shoes or gloves and in wool clothing. If caustic soda contacts the skin, wash the affected area immediately by flushing it with plenty of water to dilute and remove the caustic solution. If it gets into the eyes, flush them with water and then seek medical help promptly.

When adding dry caustic soda to the mud system—

- Stand upwind, not downwind, so the powder blows away from you. Ensure that no one is standing downwind from you.
- Wear proper protective equipment including rubber gloves, eye protection such as a full face shield, protective body wear, and a respirator.
- Put water into the chemical mixing barrel first, then slowly pour the caustic into the water in the tank. Never add water to the dry chemical because caustic soda reacts violently with water, producing heat and causing the hot solution to boil and splash.

Caustic soda mixed with water is not as hazardous as dry caustic soda because water greatly dilutes it. Organosilicate alkalizer is a substitute for caustic soda that is also strongly alkaline. Take the same precautions with it as with caustic soda.

Soda Ash

Soda ash, or sodium carbonate (Na_2CO_3), is also alkaline, though much less alkaline than caustic soda, so burns are unlikely. If a strong soda ash solution gets on your skin, flush the area with water.

Lime Hydrate

Lime hydrate ($CaOH_2$), also known as lime, slaked lime, or calcium hydroxide, is an alkaline chemical, but its fairly low solubility makes it impossible to make a strong solution. Dry lime hydrate is irritating to the lungs and skin. Be careful not to inhale it, and wash any dust off the skin.

Sodium Chromate and Sodium Dichromate

Sodium chromate and sodium dichromate are very poisonous chemicals. Handle them with as much care as caustic soda. Do not breathe them.

Bactericides

Bactericides such as paraformaldehyde and chlorinated phenols are toxic to humans as well as to bacteria. Do not inhale these chemicals, and take the same precautions when mixing them as with caustic soda. Wash your face and hands thoroughly after handling them.

Asbestos

Asbestos is very irritating to the lungs and long-term exposure can cause cancer. As with caustic soda, when adding asbestos to the mud system—

- stand upwind, not downwind, so the powder will not blow back onto you,
- pour it slowly into the mixing hopper, and
- wear gloves, goggles, and a respirator.

Oils

Some of the oils used in oil muds and to form an emulsion with water or water-base muds are irritating to the skin. Wash them off with soap and water as soon as possible. They are, of course, highly flammable, so take every precaution to keep them away from open flames.

To summarize—

Shale problems

- Unstable because of abnormal formation pressure, overburden pressure, geologic stresses, or water absorption
- Control drilling fluid properties carefully to compensate for their instability

Lost circulation

- Occurs in formations with large openings such as coarse, permeable gravels and those with caverns, fissures, or natural or induced fractures
- Can be caused by a poor cementing job
- Drill ahead using LCMs, drill blind, or use a floating mud column

High bottomhole temperature

- Increases the effect of chemicals and contaminants and accelerates the thickening of water-base muds

Abnormal formation pressure

- Too low: can cause lost circulation if the mud is too heavy
- Too high: can cause a kick if the mud is not heavy enough
- Equivalent circulating density (ECD) is a calculation of the total bottomhole pressure while circulating

- Kicks are a signal to adjust mud weight and can occur during circulation, while tripping in or out, or when circulation is shut down
- Well control involves closing in the well and weighting up the drilling mud to balance the formation pressure

Corrosion

- Occurs because oxygen, carbon dioxide, or hydrogen sulfide in drilling mud corrodes metal
- Degassers, mud additives, and specially protected pipe combat corrosion

Safety

- Wear protective clothing and goggles and avoid breathing chemical dusts and getting them on the skin
- Use proper mixing procedures and be prepared to flush the eyes and skin with plenty of water if they come into contact with dangerous materials
- Obey warnings to prevent oil fires

▼
▼
▼

Testing of Water-Base Drilling Muds

▼
▼
▼

On land rigs, the derrickhand (and on some offshore rigs the derrickhand's assistant, the pit watcher) monitors the mud for any changes in weight, viscosity, and temperature by testing, as well as changes in the size of cuttings, flow rate, and the level of mud in the tanks. A mud engineer does more sophisticated testing. Mud characteristics that the derrickhand usually measures are density, viscosity and gel strength, filtration and wall-building, and sand content. The mud engineer may test the mud's pH, liquid and solids content, the presence of contaminants, and electrolytic properties. A mud that conducts an electric current increases corrosion of the metal components in the hole. Whoever tests the mud also records the measurements in a mud record.

Two API publications, *Recommended Practice Standard Procedure for Field Testing Water-Based Drilling Fluids, 13B-1* and *Recommended Practice Standard Procedure for Field Testing Oil-Based Drilling Fluids, 13B-2*, list detailed descriptions, equipment, and procedures for testing water-base and oil muds. The information that follows elaborates on these recommended practices.

Preparing Mud Samples

To get accurate and useful results from mud tests, the samples must resemble the mud downhole, so most tests recreate some or all of the conditions in the well. For example, some tests require that the mud be stirred or agitated to simulate circulation; others require that the temperature of the mud sample be close to that which it experiences in the borehole.

Air in the mud sample can give erroneous results, so it may be necessary to add a few drops of chemical defoamer and gently stir the sample, or pour it back and forth between two containers, to get rid of the air. In stubborn cases, mud engineers may have to use a vacuum deaerator, which includes an air-tight container, vacuum lines, and a vacuum pump, to deaerate the mud sample. They put the sample in the special container, hook up a vacuum line to the container, turn on the pump, and reduce the pressure on the sample to remove the air. For details, check *RP 13-B1* and the manufacturers of vacuum deaerators.

Mud Weight (Density) Test

The crew tests the weight, or density, and viscosity of drilling mud from every 15 min to 1 hr, depending on the formation and the complexity of the mud system. Density testing is very important because it is a measure of the mud column's hydrostatic pressure. Mud density is often expressed in terms of its weight per unit of volume, although it can also be expressed in specific gravity. Most of the U.S. expresses mud weight as ppg; Canada and other parts of the world often use kg/m^3. In the western U.S., however, mud weight may be expressed in lb/ft^3. Therefore, depending on the location, the density of a 10-ppg mud can also be expressed as 1,200 kg/m^3 or 74.8 lb/ft^3. What's more, some operators require that mud density be expressed in terms of its pressure gradient, which is the amount of hydrostatic pressure a mud of a given density exerts per unit of depth. For example, a 10-ppg or 74.8-lb/ft^3 mud exerts 0.52 psi/ft, while a 1,200-kg/m^3 mud exerts 11.76 kPa/m. Often, operators prefer that pressure gradient be reported in terms of 1,000 ft or 100 m. Thus, a 10-ppg (74.8-lb/ft^3) mud's pressure gradient per 1,000 ft is $1{,}000 \times 0.52 = 520$ psi; a 1,200 kg/m^3-mud's pressure gradient per 100 m is $100 \times 11.76 = 1{,}176$ kPa.

Mud Balance

A mud balance is the usual device on the rig for measuring mud density. A mud balance is a beam balance; it works somewhat like the scale in a doctor's office (fig. 31). The balance beam has a small cup on one end that holds a precise fraction of a gallon, cubic metre, or cubic foot of mud. The beam (arm) has a sliding weight. The arm rests on a fulcrum. A level-bubble on the beam tells the operator when the beam is balanced. A graduated scale on the arm shows mud weight in ppg, kg/m^3, lb/ft^3, specific gravity, or other units. Most mud-balance arms have two graduated scales; one runs along the top and the other along the bottom. One scale shows mud weight in a given unit and the other scale in another given unit. For example, the top scale may show ppg and the bottom scale may show specific gravity or psi/1,000 ft. The mud balance weighs the precise volume of mud and is calibrated to divide the weight by the volume so that weight reads directly on the arm's scale.

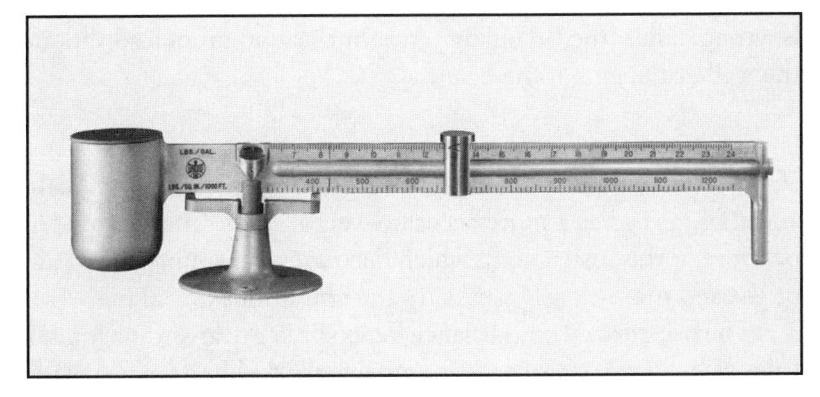

Figure 31. Mud balance

Testing Procedure

To test density, follow these steps:

1. Set the balance on a flat, level surface.
2. Measure the temperature of the sample and record it on a mud-report form. For the test to be accurate, the temperature of the mud should be between 32° and 220°F (0° and 105°C). One widely used form is the API drilling mud report.
3. Fill the clean, dry cup to the top with the mud sample, and put the lid on tightly. Some mud should ooze out of the hole in the lid to ensure that the cup is completely full.

4. Holding the lid on with a finger covering the hole in the lid, wash and dry the outside of the cup.
5. Place the beam on the base of the balance. Move the sliding weight until the arm balances—when the level-bubble is under the center line.
6. Read the mud weight at where the sliding weight points to it on the arm scale.
7. Report the mud weight to the nearest 0.1 ppg, 10 kg/m^3 (or 0.01 gram per cubic metre), or 0.5 lb/ft^3.

Calibration

Mud balances tend to get banged around on the rig so it is good practice to periodically calibrate them. To calibrate a balance, make sure the cup is clean, then fill it with clean fresh water whose temperature is 70°F (21°C) and weigh the water. The water should give a reading of 8.3 ppg, 1,000 kg/m^3, or 62.3 lb/ft^3. If the reading is wrong, adjust the balancing screw or the amount of lead shot in the well at the end of the beam.

Pressurized Mud Balance

If the mud contains entrained air or gas, a pressurized mud balance (fig. 32) gives a more accurate weight. The sample is under pressure in this instrument, which decreases the volume of the air or gas and more closely simulates the conditions downhole.

The pressurized mud balance looks similar to an ordinary mud balance but has a pressurizing pump that works like a syringe, with a plunger inside a cylinder housing.

Testing Procedure

1. Set the balance on a flat, level surface.
2. Measure the temperature of the sample and record it on a drilling mud report. The temperature should be between 32° and 220°F (0° and 105°C).
3. Fill the sample cup to about ¼ in. (6 mm) below the upper edge. Put the lid on the cup with the attached valve in the down (open) position. Push the lid down until excess mud squeezes through the hole in the lid.
4. Pull the valve up into the closed position.
5. Wash off the cup and threads with water and screw the threaded cap onto the cup.

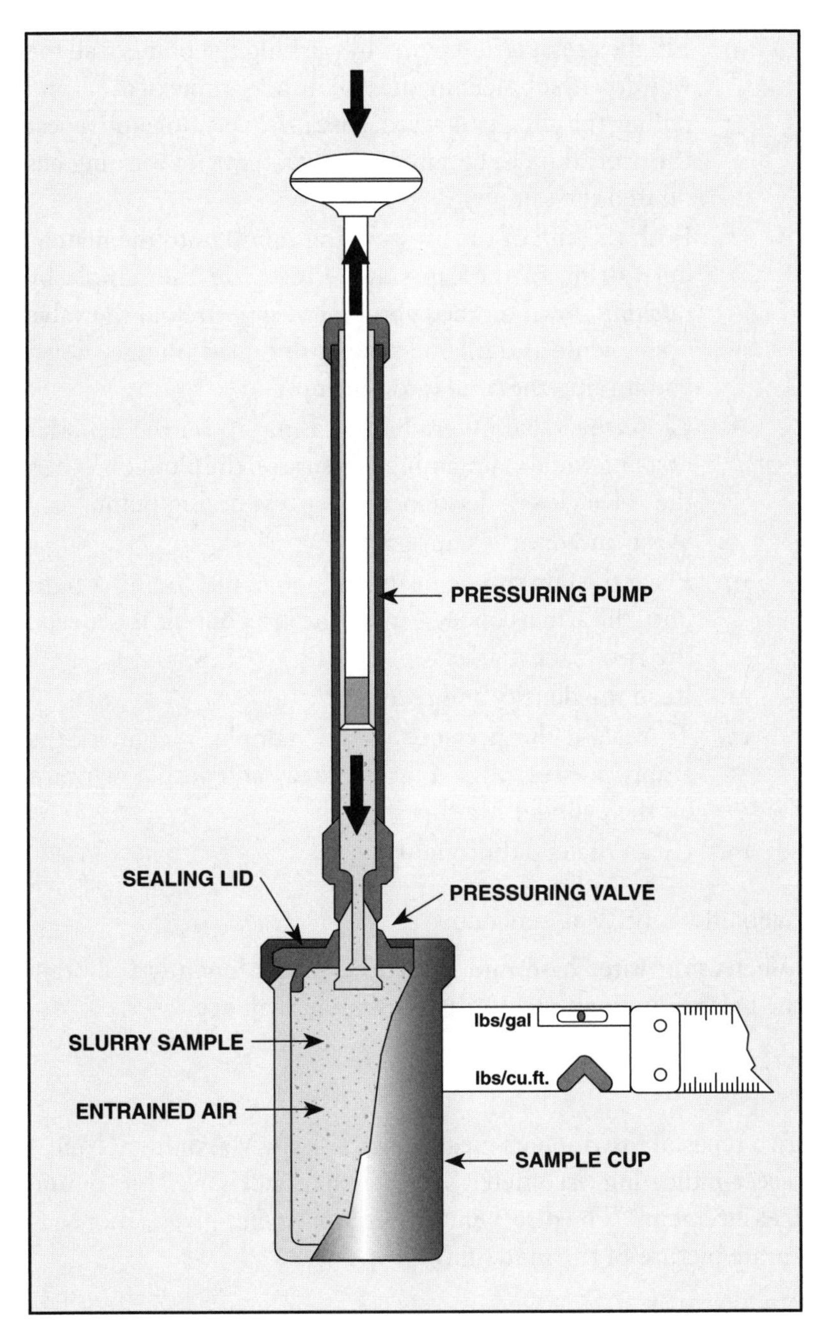

Figure 32. High-pressure mud balance

6. Fill the pressurizing pump by pushing the plunger all the way down, submersing the end into a sample of mud, and pulling the plunger upward. Discard this sample and repeat the procedure to be sure that no water from washing has diluted the sample.
7. Push the end of the pressurizing pump onto the matching O-ring of the cap valve. Pressurize the sample by pushing down on the cylinder housing to hold the valve open, while at the same time pushing the plunger down and forcing the mud into the cup.
8. Close the valve by gradually easing up on the cylinder housing while maintaining pressure on the plunger. When the valve closes, disconnect the pressurizing pump.
9. Wash and dry the cup again.
10. Place the cup on the beam and move the sliding weight until the arm balances—when the level bubble is between the two black marks.
11. Read the density and record it.
12. To release the pressure on the sample, reconnect the empty pressurizing pump to the cap and push downward on the cylinder housing.
13. Clean the cup thoroughly.

Calibration and Maintenance

When using water-base mud, grease the valve frequently. Calibrate the balance in the same way as an ordinary balance.

Viscosity and Gel Strength Tests

Two types of instruments measure viscosity, a Marsh funnel and a direct-indicating viscometer. The Marsh funnel is used for routine measurements. The direct-indicating viscometer gives a more accurate picture of the mud's flow properties.

Marsh Funnel

A Marsh funnel is a special funnel whose top is half covered with a screen. It is calibrated so that 1 quart (qt) or 1 litre (L) of clean fresh water at a temperature of 70°F (21°C) flows out of the funnel in 26 seconds (± 0.5 seconds). A cup graduated in either ounces (oz) or millilitres (mL) and with a clearly marked 1-qt (1-L) level is used to receive the mud that flows out of the funnel's opening (fig. 33). The measurement it gives is called the *funnel viscosity*. Funnel viscosity is based on how long it takes, in seconds, for 1 qt (1 L) of mud to flow through the funnel.

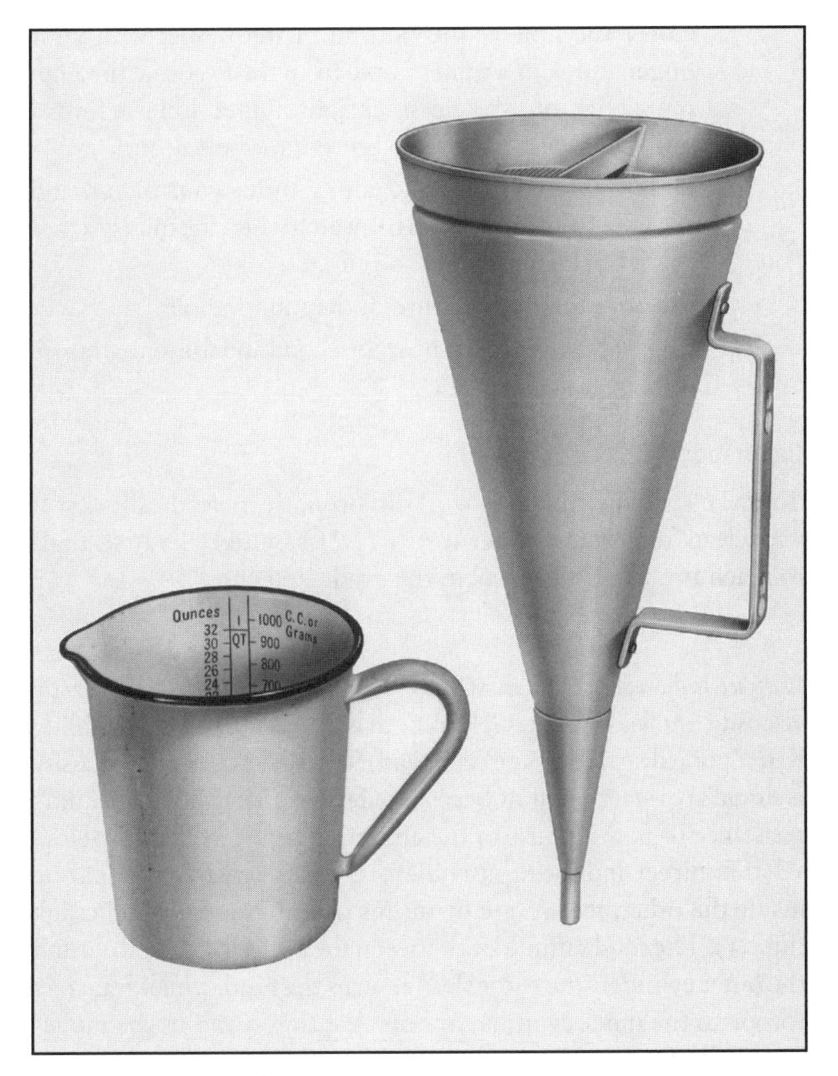

Figure 33. Marsh funnel

Funnel viscosity may be from 30 to 40 seconds for low-solids muds, from 40 to 50 seconds for high-solids muds, and above 50 seconds for heavier-weight muds. Funnel viscosity of an oil mud is about the same as a water-base mud, but the temperature of the sample is more important because the viscosity of oil varies with temperature more than that of water.

Testing Procedure

This simple test requires the Marsh funnel, a graduated cup to catch the mud sample, a mud sample, a stopwatch, and a thermometer.

1. Cover the hole at the bottom of the funnel with your finger and pour a mud sample from the flow line through the screen into the clean, upright funnel. Fill the funnel until the fluid reaches the bottom of the screen.
2. Hold the funnel over the cup, remove your finger, and start the stopwatch. Stop the watch when the mud reaches the 1-qt (1-L) mark on the cup.
3. Measure the temperature of the mud sample.
4. Report the time to the nearest second and the temperature to the nearest degree.

Calibration

To make sure the funnel is working properly, periodically test it with clean fresh water. Water at 70°F (21°C) should take 26 seconds to reach the 1-qt (1-L) level in the graduated cup.

Direct-Indicating Viscometer

A *direct-indicating viscometer* (fig. 34) measures gel strength, plastic viscosity, and yield point. Gel strength is a measure of a fluid's ability to temporarily gel (become semisolid) when at rest. Plastic viscosity is a fluid's resistance to flow because of friction. Yield point is a fluid's resistance to flow because of the attraction between clay particles.

The direct-indicating viscometer consists of two cylinders, one inside the other, that rotate by means of a motor or a hand crank (fig. 35). The mud sample sits between the two cylinders. Rotating the outer cylinder (the rotor sleeve) turns the mud, which transfers torque to the inner cylinder, or bob. A spring restrains the movement of the bob, and a dial indicates how far the spring moves. In other words, how well the mud transfers torque determines the amount of movement of the bob spring. The manufacturer adjusts the viscometer to give plastic viscosity and yield point readings at rotor sleeve speeds of 300 and 600 revolutions per minute (rpm).

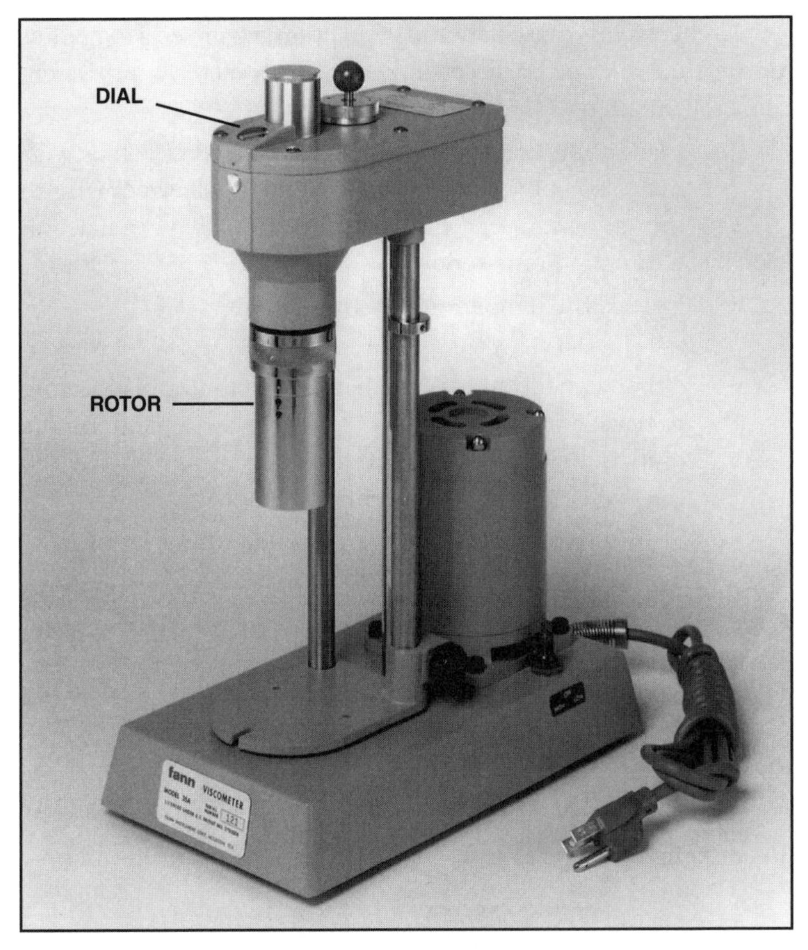

Figure 34. *Direct-indicating viscometer*

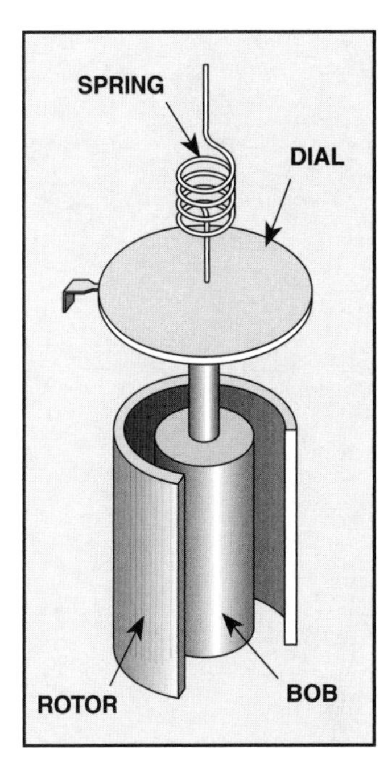

Figure 35. Diagram of a direct-indicating viscometer. Read the deflection of the bob in degrees from a scale on the dial.

Testing Procedures for Plastic Viscosity and Yield Point

Temperature of the mud sample is crucial when testing in a direct-indicating viscometer. Take the measurements while the mud is within 10°F (6°C) of the temperature at the place of sampling. Measuring within 5 minutes of removing the sample from the mud stream helps keep the temperature in the correct range.

One caution to keep in mind when using a viscometer: the maximum operating temperature of the instrument is normally 200°F (93°C). If the mud is above this temperature, be sure to use a viscometer with a solid metal bob or, if using one with a hollow bob, make sure that it is completely dry inside. Liquid trapped inside a hollow bob may vaporize when immersed in mud that is too hot and cause the bob to explode. If you don't know whether your viscometer has a solid or hollow bob, check with the manufacturer.

To check plastic viscosity and yield point, you need a container for the mud sample, the direct-indicating viscometer, a stopwatch, and a thermometer.

1. Place a mud sample into the container that comes with the viscometer and immerse the rotor sleeve exactly to the scribed line on the sleeve. Write down the place of sampling on the report.
2. Record the temperature of the sample.
3. Set the sleeve to rotate at 600 rpm and begin rotating it, either with the motor or by hand. Wait for the dial reading to reach a steady value (how long it takes depends on the mud, but it is usually a matter of a few minutes). Record the dial reading for 600 rpm.
4. Set the sleeve to rotate at 300 rpm, rotate it, and wait for the dial to become steady. Record the reading for 300 rpm.

To determine plastic viscosity in centipoises (cp), use the following equation—

$$PV_{cp} = 600\ rpm\ reading - 300\ rpm\ reading \qquad \textit{Equation 6}$$

where

PV_{cp} = plastic viscosity, cp.

To determine yield point in lb/100ft², use the following equation—

$$YP_{lb/100ft2} = 300\ rpm\ reading - plastic\ viscosity \qquad \textit{Equation 7}$$

where

$YP_{lb/100ft2}$ = yield point, lb/100ft².

If using SI units, and the viscometer is not calibrated in SI units, simply multiply the yield point in lb/100ft² by 0.48 to convert to Pa.

Testing for Gel Strength

To measure gel strength requires two tests on the same mud sample at the same temperature. The first, the 30-second test, shows how quickly the mud gels. The second, the 10-minute test, shows how strongly it gels.

1. In the same container as used in previous tests, lower the direct-indicating viscometer into the mud sample, and stir the sample by hand or motor for 10 seconds at high speed.
2. Let the mud stand undisturbed for 10 seconds and set the motor to turn the cylinder very slowly, at 3 rpm (or turn the hand crank slowly and steadily to produce a positive dial reading.)

3. Take a reading from the dial when it reaches a maximum value. Record the dial reading as the initial gel strength, in lb/100 ft^2 (Pa).
4. Restir the sample for 10 seconds at high speed.
5. Let the mud stand undisturbed for 10 minutes this time, and then set the motor to turn the cylinder very slowly, at 3 rpm (or turn the hand crank slowly and steadily to produce a positive dial reading).
6. Take a reading from the dial when it reaches a maximum value. Record the dial reading as the 10-minute gel strength, in lb/100 ft^2 (Pa).

Calibration

Calibrate a viscometer by using a certified calibration fluid. A certified calibration fluid has a standard known viscosity and is commercially available in different ranges corresponding to commonly used drilling mud viscosities, such as 50 cp or 100 cp. Run the fluid through a clean, dry viscometer as if conducting an ordinary plastic viscosity test at room temperature, and compare the test viscosity to the certified viscosity of the fluid. The viscometer should be accurate to 1.5 units. Be careful not to contaminate the calibration fluid with water. Calibrate a new viscometer before using it, and then check it monthly thereafter while it is in service.

Filtration and Wall Building Tests

Filtration tests measure the relative amount of liquid in the mud that migrates into a permeable formation and the thickness of the filter cake deposited on the walls of the hole. The test can be made at low or high temperature and pressure. An instrument called a *filter press* measures filtration rate.

Low-Temperature, Low-Pressure Filter Press

The simplest type of filter press tests mud at low temperatures and pressures. It consists of a cell for the mud at the top of the press, filter paper, and a graduated cylinder below to catch the filtrate and measure it in cubic centimetres (cm^3). The entire assembly is on a stand. A CO_2 or nitrogen cartridge supplies the pressure needed, and a timer measures 30-second intervals. Use one thickness of the proper 90-mm filter paper, Whatman No. 50, S&S No. 576, or the equivalent.

Testing Procedures

1. Make sure the cell is clean and dry, especially the screen in the bottom, and that the gaskets are not distorted or worn.
2. Pour a mud sample into the cell to within 0.5 inch (13 mm) of the top. Measure and record the temperature of the mud.
3. Assemble the cell, pressure cartridge, and filter paper, and place the graduated cylinder under the drain tube.
4. Close the relief valve, and adjust the pressure regulator so that a pressure of 100 ± 5 psi (690 ± 35 kilopascals) is applied in 30 seconds or less. The testing begins when you apply pressure.
5. After 30 min, read on the graduated cylinder the amount of liquid that has filtered out of the mud. Shut off the pressure regulator and open the relief valve slowly. Report the time if it is different from 30 min.
6. Report the volume of filtrate in cm^3 to the nearest 0.1 cm^3 as the API filtrate, and the initial mud temperature in °F (°C). Save the filtrate for chemical testing.
7. Remove the cell from the stand (make sure all pressure has bled off). Being extremely careful not to disturb the filter cake, save the filter paper. Disassemble the cell and discard the mud. Gently wash the filter cake on the paper in a stream of water.
8. Measure and record the thickness of the filter cake to the nearest 1⁄32 inch (0.8 mm).
9. Describe the quality of the filter cake, using such terms as hard, soft, tough, rubbery, and so on.

Calibration

Gaskets or O-rings must have an inside diameter between 2.99 and 3.03 in. (75.8 to 76.9 mm). Check the gaskets or O-rings with a gauge, and do not use them if they do not meet the tolerance.

High-Temperature, High-Pressure Filter Press

High-temperature tests are run at the actual bottomhole temperature and pressure. A filter press for this purpose (fig. 36) consists of

- a pressure source (CO_2 or nitrogen),
- pressure regulators,
- a cell that can withstand pressures up to 1,300 psi (8,970 kPa),
- a system for heating the cell,
- a pressurized collection cell that maintains proper back pressure, and
- a stand.

Other equipment needed for the test include a timer, a thermometer, a graduated cylinder, and a high-speed mixer.

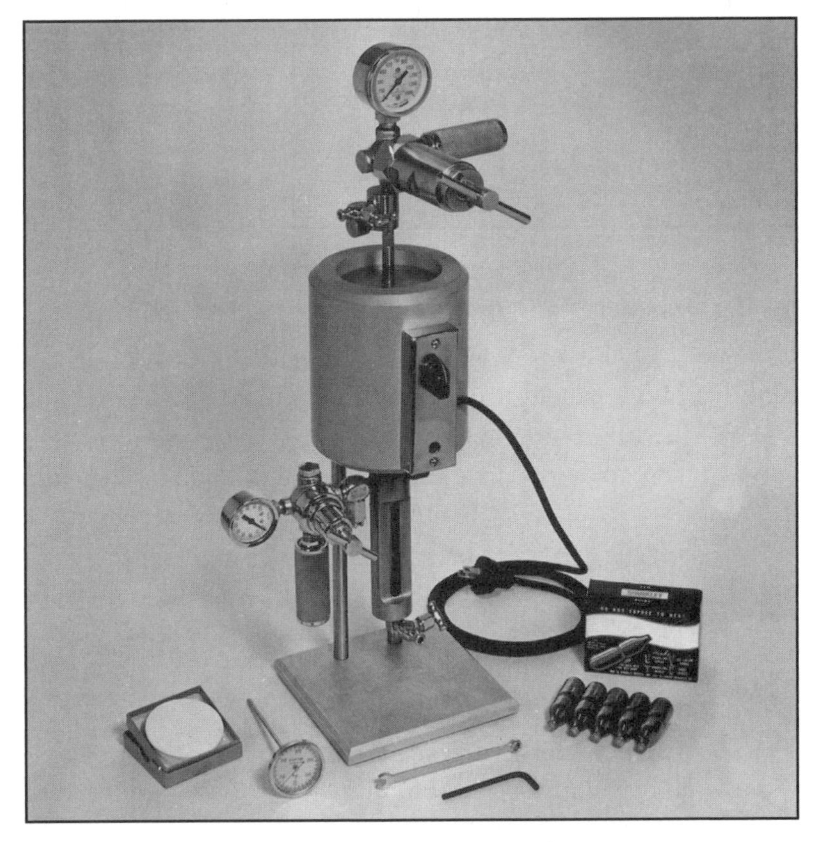

Figure 36. High-pressure, high-temperature filter press

Table 12 shows the recommended minimum back-pressure needed to prevent the filtrate from vaporizing or evaporating. The table only gives values normally found in field testing. In a laboratory, higher temperatures and pressures may be used. In any case, never exceed the manufacturer's recommendations for sample volumes, temperatures, and pressures.

Table 12
Minimum Back-Pressure Required to Prevent Filtrate Vaporization or Evaporation

Test Temperature		Vapor Pressure		Minimum Back-Pressure	
°F	°C	psi	kPa	psi	kPa
212	100	14.7	101	100	690
250	121	30.0	207	100	690
300	149	67.0	462	100	690

Testing Procedures—Temperatures up to 300°F (149°C)

Use one thickness of Whatman No. 50, S&S No. 576, or the equivalent filter paper for temperatures up to 400°F (260°C).

1. Place the thermometer in the well in the jacket and preheat it to 10°F (6°C) above the desired temperature. Adjust the thermostat to maintain this temperature.
2. Stir the mud sample for 10 min with a high-speed mixer. Then pour it into the cell, to within 0.5 in. (13 mm) from the top. Do not fill any higher than this to allow for the mud to expand as it heats.
3. Install the filter paper.
4. Assemble the cell, close both the top and bottom valves, and put it into the heating jacket. Move the thermometer to the well in the mud cell.
5. Connect the high-pressure collection cell to the bottom valve and lock it in place.
6. Connect the pressure source to the top valve and lock it in place.

7. Keeping the valves closed, adjust the top and bottom regulators to 100 psi (690 kPa). Open the top valve to apply 100 psi (690 kPa) to the mud sample. Maintain this pressure until the temperature stabilizes at the desired heat. Never heat the sample for more than 1 hour.
8. When the sample reaches the desired temperature, increase the pressure at the top to 600 psi (4,140 kPa), and open the bottom valve to start filtration.
9. Collect the filtrate for 30 min, keeping the temperature within ± 5°F (± 3°C). If the back-pressure rises above 100 psi (690 kPa) during the test, cautiously reduce the pressure by drawing off some of the filtrate.
10. Record the total volume, temperature, pressure, and time.
11. Close the top and bottom valves. Bleed off the pressure from the regulators. Keep the cell upright and let it cool to room temperature.
12. The pressure will still be about 500 psi (3,450 kPa), so bleed it again before disassembling. Then make sure both valves are tightly closed and all pressure is off the regulators. Remove the cell from the jacket, being very careful to save the filter paper. Place the cell upright, open the valve to bleed off pressure from the cell, and open it. Discard the mud sample and retrieve the filter paper. Wash the filter cake gently in a stream of water.
13. Measure and report the thickness of the filter cake to the nearest 1⁄32 inch (0.8 mm).

Testing Procedures—Temperatures above 300°F (149°C)

Testing at high temperatures and high pressures calls for extra safety precautions. Not all filter presses are capable of testing above 300°F (149°C), so check the rating of the unit you are using. In addition, make sure that all pressure cells have manual relief valves, and the heating jacket has both an overheat safety fuse and a thermostatic cutoff.

A difference at temperatures higher than 400° F (260°C) is that you use a Dynalloy X-5 or equivalent porous disc instead of filter paper.

1. Place the thermometer in the well in the jacket and preheat it to 10°F (6°C) above the desired temperature. Adjust the thermostat to maintain this temperature.

2. Stir the mud sample for 10 min with a high-speed mixer. Then pour it into the cell, to within 0.5 in. (13 mm) from the top. Do not fill any higher than this to allow for the mud to expand as it heats.
3. Install the filter paper or disc.
4. Assemble the cell, close both the top and bottom valves, and put it into the heating jacket. Move the thermometer to the well in the mud cell.
5. Connect the high-pressure collection cell to the bottom valve and lock it in place.
6. Connect the pressure source to the top valve and lock it in place.
7. Keeping the valves closed, apply the recommended back-pressure (see Table 10) for the test temperature to both top and bottom. Open the top valve, applying the same pressure to the mud while heating. Maintain this pressure until the test temperature is reached and stabilized. Never heat the sample for more than 1 hr.
8. When the sample reaches the desired temperature, increase the pressure at the top by 500 psi (3,450 kPa) over the back-pressure being held, and open the bottom valve to start filtration.
9. Collect the filtrate for 30 min, keeping the temperature within ± 5°F (± 3°C) and maintaining the proper back-pressure. If the back-pressure begins to rise during the test, cautiously reduce the pressure by drawing off some of the filtrate.
10. Record the total volume, temperature, pressure, and time.
11. Close the top and bottom valves. Bleed off the pressure from the regulators. Keep the cell upright and let it cool to room temperature.
12. The pressure will still be about 500 psi (3,450 kPa), so bleed it again before disassembling. Then make sure both valves are tightly closed and all pressure is off the regulators. Remove the cell from the jacket, being very careful to save the filter paper. Place the cell upright, open the valve to bleed off pressure from the cell, and open it. Discard the mud sample and retrieve the filter paper. Wash the filter cake gently in a stream of water.
13. Measure and report the thickness of the filter cake to the nearest 1/32 inch (0.8 mm).

Measuring Sand Content

The sand content is a measurement of the amount of particles in a mud that are larger than 74 μ. A sand test measures only the size of the particles, not what they are made of.

A screen set (fig. 37) is the instrument for measuring sand content, in percent of sand per volume of mud. It has a 200-mesh sieve inside a collar which fits into a small funnel and a measuring tube with marks that indicate from 0 to 20 percent sand.

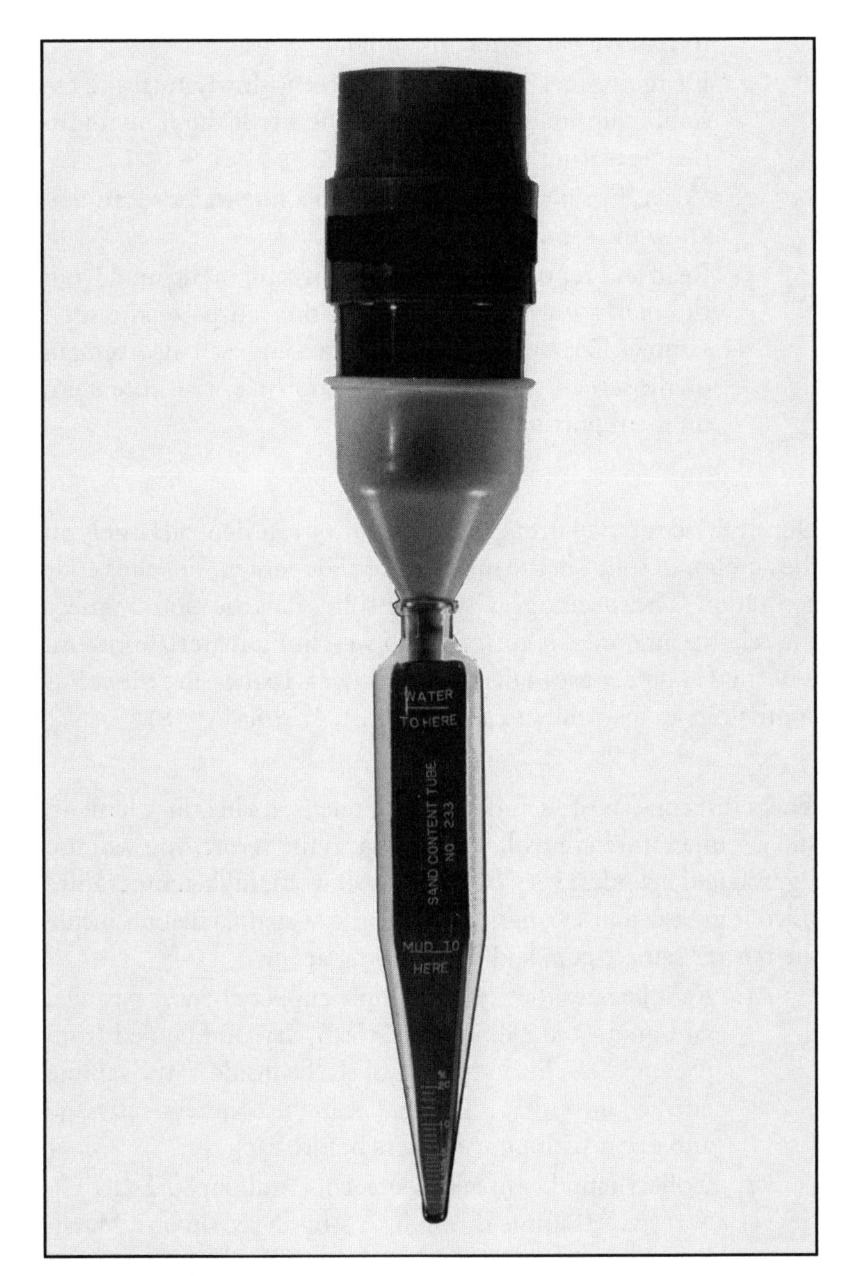

Figure 37. Screen set

Testing Procedure

1. Fill the measuring tube with a mud sample to the "mud" mark. Add water to the "water" mark. Close the tube and shake it vigorously.
2. Pour the shaken mixture onto the clean, wet screen. Discard the liquid that passes through the screen. Add more water to the measuring tube. Shake it again and pour it onto the screen again. Repeat this process until the tube is clean. Wash the sand that stays on the screen to remove any remaining mud.
3. Fit the funnel on top of the screen, slowly turn the assembly upside down, and put the tip of the funnel into the measuring tube.
4. Wash the sand into the tube with a fine water spray, and allow the sand to settle out.
5. Read and report the percentage of sand in the mud from the marks on the tube. Report the source of the mud sample. Coarse solids other than sand will also remain on the screen, such as lost circulation material; note these on the reporting form as well.

Solids, Water and Oil Content

Density, viscosity, gel strength, and filtration rate depend largely on the amount of solids in the mud—the solids content or solids concentration. The specific gravity of the solids tells the mud engineer the relative amounts of cuttings and weighting material present. The mud engineer uses a distillation process to measure the solids content, in an instrument called a retort, or still (fig. 38).

Testing Procedure

The retort consists of a sample cup, a condenser, a heating element, and a temperature control. In addition to the retort, you will use a graduated cylinder; very fine, 000 steel wool; high-temperature silicone grease; pipe cleaners; a putty knife or spatula that fits inside the retort's sample cup; and a defoaming agent.

1. First be sure that retort sample cup, condenser passage, and graduated cylinder are clean, dry, and cooled from previous use. Periodically polish the inside of the sample cup and lid with steel wool. Clean the condenser passage and dry it with pipe cleaners before each test.
2. Collect a mud sample and let it cool to about 80°F (26°C). Screen the sample through a 20-mesh screen on a Marsh funnel to remove large solids, such as lost circulation material or large cuttings.

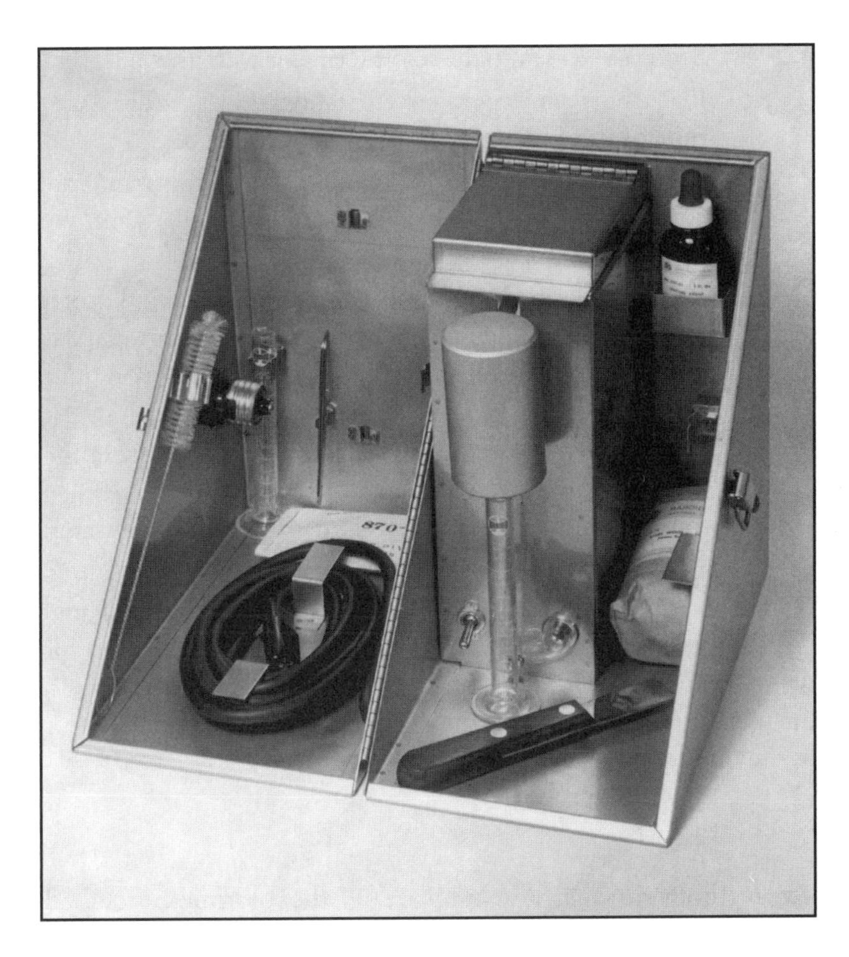

Figure 38. Mud still

3. If the mud sample contains air or gas, add 2 to 3 drops of defoaming agent to about 300 cm^3 of mud. Stir the sample slowly to release the gases.
4. Lubricate the threads of the sample cup and condenser tube with a light coating of silicone grease. This prevents vapor loss through the threads and also makes unscrewing them and cleaning them at the end of the test easier.
5. Lightly pack a ring of steel wool into the chamber above the sample cup. Experience will teach you how much to use to keep the solids from boiling over into the graduated cylinder.
6. Fill the sample cup with the prepared mud sample.
7. Put the lid on the cup and allow mud to overflow through the hole in the lid. Hold the lid in place and wipe off the overflow from the cup and lid. Make sure grease still covers the threads and that the hole in the lid is not plugged up.

8. Screw the cup onto the retort chamber with its condenser.
9. Place the clean, dry graduated cylinder under the discharge tube of the condenser.
10. Heat the retort. You will see liquid falling from the condenser. Continue heating for 10 minutes after the condensate stops falling.
11. Remove the graduated cylinder containing liquid. Look for solids in the liquid—if there are any, this means that the mud boiled over from the sample cup and you must repeat the test.
12. Read and record the volumes of water and oil from the marks on the graduated cylinder after the liquid has cooled to room temperature. The oil floats on top of the water, so reading the two volumes is easy.

Subtracting this measurement from the original volume of mud gives the solids content. Saltwater muds require a correction for the salt content. After calculating the solids, water, and oil content, the mud engineer compares these values with the actual behavior of the mud in drilling.

Determining pH

Two methods are available to measure pH: the colorimetric method and the electrometric method. The colorimetric method uses chemically treated paper strips, (like litmus paper). The electrometric method uses a pH meter (fig. 39) for a very accurate value. It works on the principle that when a thin membrane of glass separates two solutions with different pH values, an electrical current develops that can be amplified and measured. This type of test is not normally done on the rig but in a lab.

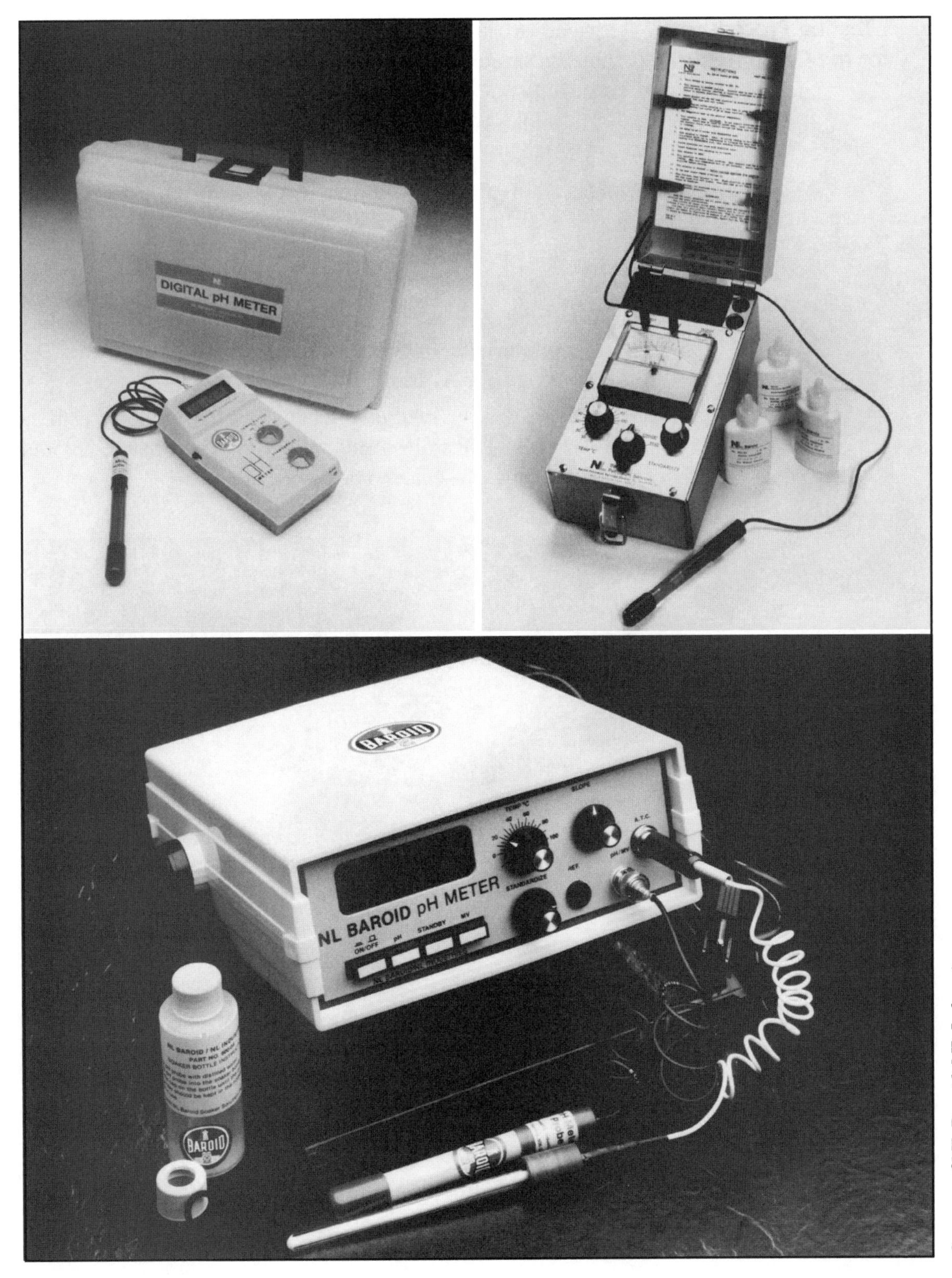

Courtesy of NL Baroid, NL Industries

Figure 39. pH meter

Testing Procedure for the Colorimetric Method

The paper used in pH testing in the field is called pHydrion paper. It comes in several pH ranges, on rolls in dispensers (fig. 40).

1. Choose the pH range that you expect the mud to be within, and remove a 1-in. (25-mm) strip of paper from the dispenser. Place it gently on the surface of the mud.
2. Allow a few seconds to a minute or two for the paper to soak up the filtrate and change color.
3. Match the color of the strip with the chart on the side of the dispenser, read and record the pH.
4. If the color of the strip does not match any of the colors on the chart, repeat the test with a strip from a different range of pHydrion.

Be aware that this method is reliable only in simple water-base muds. Mud solids, dissolved salts and chemicals, and dark-colored liquids can cause the color to be inaccurate.

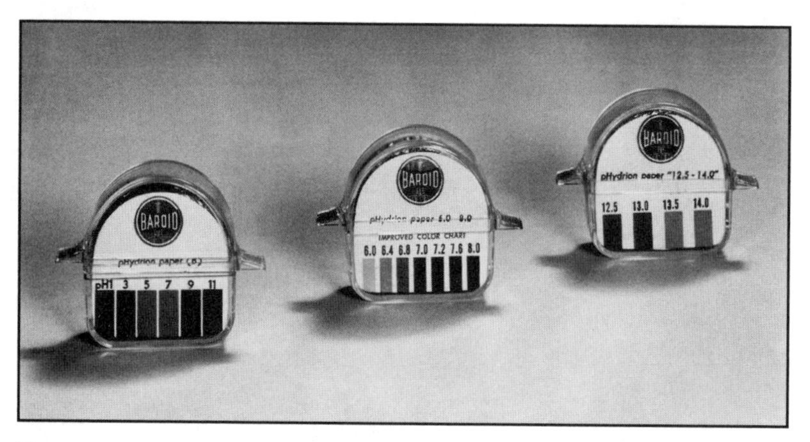

Figure 40. Rolls of pHydrion paper

Methylene Blue Capacity

Methylene blue is a dye that colors the reactive clays (bentonite and drilled clay solids) in a mud sample. This test is used to calculate the percentage of clay in a sample by volume.

Testing Procedure

Since muds often contain other substances that also absorb methylene blue, such as lignosulfonates and lignites, for example, pretreat the sample with hydrogen peroxide to remove their effect on the test.

A procedure called titration allows measurement of the amount of methylene blue added to a precise amount of drilling fluid. In titration, you add a solution (in this case, the methylene blue) drop by drop to another solution to get a response—a color change or a precipitate, for example. The change is the endpoint.

Use an Erlenmeyer flask, syringe, burette, graduated cylinder, stirring rod, hot plate, and Whatman No. 1 filter paper (fig. 41).

1. Stir or defoam the drilling fluid sample to release entrained air or gas before testing it. Use the syringe to measure and add 2 cm^3 (or 2 mL) of sample to 10 cm^3 (mL) water in the Erlenmeyer flask.
2. Add 15 cm^3 (mL) of a 3 percent hydrogen peroxide solution and 0.5 cm^3 (mL) of dilute sulfuric acid. Boil gently for 10 min, without boiling all the moisture out.
3. Dilute to about 50 cm^3 (mL) with water.
4. Add methylene blue to the flask , 0.5 cm^3 (mL) at a time. After each addition, swirl the liquid in the flask for about 30 seconds. While the solids are still suspended, remove a drop of fluid with the stirring rod and place on the filter paper. Stop when the dye appears as a blue or turquoise ring around the dyed solids.

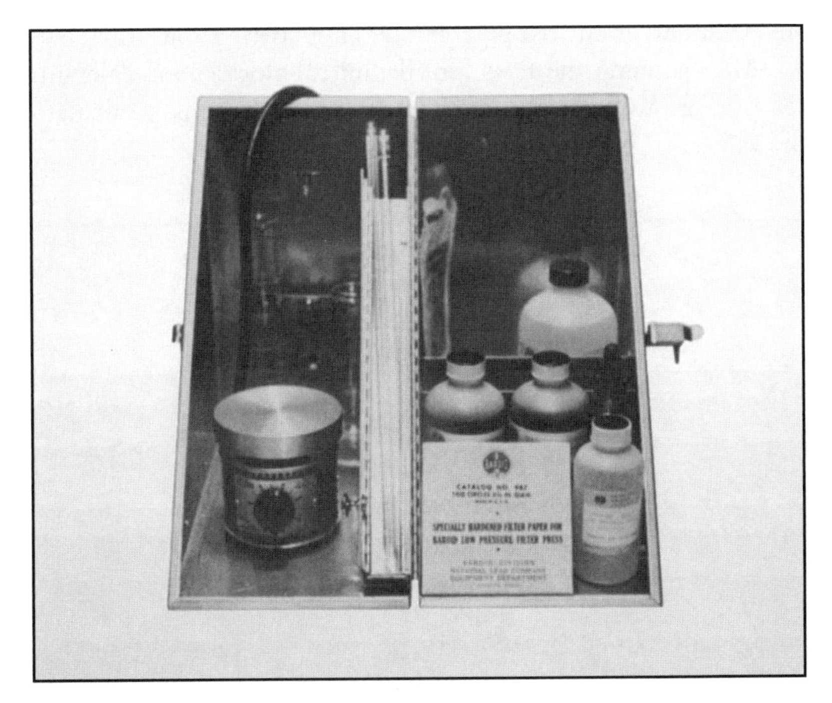

Figure 41. Methylene blue test kit

Shake the flask for 2 more min and place another drop on the filter paper. If it again forms a blue ring, you are finished. If it does not, continue adding methylene blue until a drop taken after 2 min shows the blue color.

Calculation

Report the methylene blue capacity by dividing the amount of methylene blue needed to reach the endpoint by the amount of drilling fluid in the sample.

$$MBT_{cm3/cm3} = \frac{methylene\ blue,\ cm^3}{drilling\ fluid,\ cm^3} \qquad \text{Equation 8}$$

where

$MBT_{cm3/cm3}$ = methylene blue capacity, cm^3/cm^3

Chemical Analysis

Some chemical tests on the mud filtrate determine the presence of contaminants, such as salt or anhydrite, or determine extreme alkalinity in high-pH muds. The same tests can be run on makeup water, which in some areas contains dissolved salts that affect mud treatment.

A filtrate analysis kit (fig. 42) contains the equipment usually needed for these types of tests: an automatic burette, reagent bottle, dropper bottle, casserole (a round, handled dish for heating), graduated cylinder, graduated pipette, and glass stirring rod.

After running the tests for alkalinity, chloride, and calcium, refer to API *Recommended Practice 13B-1* for equations to calculate the values.

Figure 42. Filtrate analysis kit

Determining Extreme Alkalinity

Because a change in one number value on the pH scale represents 10 times the previous value, the alkalinity of a high-pH mud can vary considerably with no measurable change in pH. In these fluids, filtrate analysis is a better measure of alkalinity than a pH test.

Testing Procedure for Filtrate Alkalinity (P_f)

A process called titration is the method for analyzing filtrate alkalinity (P_f).

1. Measure 1 or more cm^3 of filtrate into a titration vessel, and record the volume.
2. Add 2 to 3 drops of phenolphthalein indicator solution. If the filtrate stays the same color, P_f is 0. If the solution looks pink, continue the test.
3. Add a 0.02 normal (N) solution of sulfuric acid drop by drop from an automatic burette or a pipette to the filtrate, stirring continuously, until the sample changes from pink to colorless. The endpoint of titration occurs when the solution changes color. Record the volume of sulfuric acid (in cm^3) you added.
4. Add 2 to 3 drops of methyl orange indicator solution. The filtrate will turn orange.
5. Add 0.02N sulfuric acid drop by drop from the pipette until the color of the solution changes from orange to pink (another endpoint). Record the volume of acid you added.

Obviously, this test depends on the filtrate being able to show the pink and orange colors. If it is too dark or colored for this, use a pH-meter on the sample and record the first endpoint when the pH reaches 8.3 and the second when the pH reaches 4.3.

Salt Concentration (Chloride) Test

The salt or chloride ion (Cl^{-1}) test measures the chloride ion concentration in the mud filtrate. If the chloride concentration changes during drilling, it may be an indication that salt water from a formation has entered the hole. The test is also useful in areas where salt can contaminate the drilling fluid. When the salt concentration is greater than 6,000 ppm, the mud engineer may decide to change the mud treatment program.

Testing Procedure

For this test, use graduated pipettes, a titration vessel (preferably white), and a stirring rod.

1. Measure 1 or more cm^3 of filtrate into a titration vessel.
2. Add 2 to 3 drops of phenolphthalein indicator. If the solution turns pink, add 0.02N sulfuric acid drop by drop through a pipette while stirring until the color disappears. If the filtrate is deeply colored, add an extra 2 cm^3 of sulfuric acid and stir. Then add 1 gram (g) of calcium carbonate and stir.
3. Add 25 cm^3 of the filtrate solution to 50 cm^3 of distilled water and 5 to 10 drops of potassium chromate indicator.
4. Stir continuously while adding standard silver nitrate solution drop by drop from a pipette until the color changes from yellow to orange-red and persists for 30 seconds.
5. Record the amount of silver nitrate solution you used to reach the endpoint. If you used more than 10 cm^3 of silver nitrate solution, repeat the test with a smaller sample of filtrate.

Hardness and Calcium Concentration Test

Hard water is water that contains dissolved calcium and magnesium salts. We notice that water is hard when it is difficult to produce a lather in it with soap. Drilling clays do not work as well in hard water—the harder the water, the more bentonite it takes to make a good gel mud.

Experienced oilfield workers are familiar with "gyppy," or gypsum-bearing formations, such as anhydrite. Gypsum is the naturally occurring form of calcium sulfate. Calcium salts may also enter the drilling mud by drilling through cement plugs, or limy shales. Too much calcium contamination causes high fluid loss and a fast gelling.

Testing Procedure

You will need a 150-cm^3 beaker, graduated pipettes, a hot plate, pH strip, and the following solutions: EDTA, a buffer solution, hardness indicator solution, acetic acid, masking agent, sodium hypochlorite, and deionized or distilled water.

1. Measure 1 or more cm^3 of the water or filtrate sample into the beaker. If the sample is colorless or only lightly colored, go to step 5.
2. If the sample is colored, add 10 cm^3 sodium hypochlorite and stir.
3. Add 1 cm^3 acetic acid and stir.
4. Boil the sample for 5 minutes. Add deionized or distilled water as needed to keep the volume up. Immerse a pH strip in the sample. If the paper is bleached white, continue boiling. When it does not change color, cool the sample and wash the sides of the beaker with deionized or distilled water.
5. Dilute the sample to 50 cm^3 with deionized or distilled water. Add about 2 cm^3 of buffer solution and swirl to mix.
6. Add 2 to 6 drops of hardness indicator and stir. The solution will turn wine-red if calcium or magnesium is present.
7. Add EDTA drop by drop while stirring until the solution turns blue. The endpoint is reached when adding more EDTA produces no further red-to-blue color change.

Other Chemical Tests

Mud engineers may do a number of other more sophisticated tests if necessary to determine the presence of contaminants or to help control mud properties. They can do the same tests on the water used to make up the mud. Some of these tests are for calcium, calcium sulfate, formaldehyde, sulfide, carbonate, and potassium.

To summarize—

Main tests done on site

- Weight (density) with a mud balance
- Viscosity and gel strength with a Marsh funnel or direct-indicating viscometer
- Filtration and wall-building capacity with a filter press
- Sand content with a screen set
- Solids, water, and oil content with a mud still
- pH with litmus paper or pH meter
- Percentage of clay with methylene blue dye
- Contaminants with a filtrate analysis kit

▼
▼
▼

Testing of Oil Muds

▼
▼
▼

Generally, the tests for oil-base and invert-emulsion muds are the same as those for water-base muds. Some, however, take on a special importance in oil muds that they do not have for water-base muds. Others are exclusively for oil muds.

Funnel Viscosity

The funnel viscosity of an oil mud as measured at the flow line should be about the same as that of a good water-base mud used for the same purpose. But one difference between an oil mud and a water-base mud is important: temperature affects the funnel viscosity of an oil mud more than a water-base mud because the viscosity of oil varies more with temperature than water. A drastic drop in atmospheric temperature can reduce the flow-line temperature of an oil mud enough to cause a significant increase in the funnel viscosity, although its downhole viscosity may remain unchanged. Because of this effect, API procedure recommends that reports include mud temperature along with funnel viscosity, especially for oil muds.

API high-temperature, high-pressure filtration tests attempt to simulate downhole temperature and pressure but in many cases do not reach the actual downhole pressures in many wells. In deep wells, for example, hydrostatic pressure may be 10,000 to 20,000 psi (68,900 to 138,000 kPa). Such high pressures increase the viscosity and the downhole filtration rate of oil muds. Water and water-base muds are less affected by such high pressures.

Water Content

Water content, percent of water by volume, may be measured with either a water-determination apparatus from the American Society for Testing and Materials (ASTM) or a field model electric retort, which distills the mud and also gives the oil and solids content.

Water is a contaminant for oil muds, so it is important to maintain field conditions that protect the mud system against rain or other sources of extraneous water. Large amounts of water may thicken oil muds. Also, excess water may require additional emulsifiers and chemicals to maintain a tight emulsion and prevent weighting particles, such as barite, from becoming water wet.

Stability Test

An oil mud is subject to instability because of the extreme conditions in wells where it is often used. The emulsion can break down, causing the oil phase and the solids to separate out.

Electrical Stability

The electrical stability (ES) of an oil mud is related to the stability of the emulsion. The test measures the stability in volts. A weighted oil mud should have an ES above 480 volts. A low-weight oil mud's ES is 200 volts or higher. If the ES is low and too much water is not the problem, minor chemical treatment can restore the mud. But if ES is low and the water content is too high, the mud engineer may have to completely rebuild the mud with additives.

Testing Procedure

The equipment for testing electrical stability is a meter that measures a current passing through the mud. The meter is connected to two electrodes inside a probe that conduct a voltage-ramped, sinusoidal electrical signal through the mud. You will also need a thermometer, a 12-mesh screen or Marsh funnel, and a thermostatically controlled viscometer cup.

1. After inspecting and calibrating the equipment (covered later), screen the mud sample and place it into the viscometer cup maintained at 120 ± 5°F (50 ± 2°C). Measure and record the mud's temperature.
2. Clean the electrode probe by wiping it with a paper towel. Pass the towel through the electrode gap a few times. Swirl the probe in a sample of the oil used to formulate the mud. If none is available, use another oil or a solvent such as isopropanol. Do not use a detergent or aromatic solvent such as xylene. Clean and dry the probe.

3. Stir the 120°F (50°C) sample by hand with the electrode probe for about 10 seconds. Set the electrode probe in the sample so that it does not touch the bottom or sides of the container, and be sure that mud completely covers the probe.
4. Follow the procedure in the meter's operating manual to begin the voltage ramp test. Do not disturb the electrode during the voltage ramp test.
5. At the end of the test, record the ES value displayed on the readout.
6. Repeat steps 1–5 with the same mud sample. The two ES values should be within 5 percent of each other. If they are not, check the meter and electrode to make sure they are working properly.
7. Take the average of the two readouts and report them.

Calibration and Maintenance

Inspect the electrode probe for damage and make sure it is free of deposits. The connector to the meter should be clean and dry. Disconnect the electrode probe, if possible, and run a voltage ramp test according to the operating meter's manual. Reconnect the probe and repeat the test in air. For both tests, the ES reading should reach the maximum allowed by the meter. If it does not, clean or replace the probe or connector.

Repeat the voltage ramp test with the probe sitting in tap water. The ES reading should be 3 volts or less. Clean or replace the probe if it is over 3 volts.

The meter comes with two calibration resistors or diodes to test the accuracy of the voltage readouts. If the ES readings with these calibration diodes are outside the expected value, return the meter to the supplier for adjustment or repair.

Separation Tests

Aging a mud sample at bottomhole temperature and pressure determines whether the oil mud will be stable under these conditions. After cooling and depressurizing the sample but before stirring it, observe whether free oil (oil that is not emulsified) is present on the top of the sample. If it is present, measure top-oil separation in 1⁄32 in. (mm). Next observe whether the solids settle from the mud when hot, and if so, whether they form a hard or soft layer. The condition of the solids tells you the effect of temperature on the filtration and the type of filter cake to expect. A harder cake will stabilize better.

Chemical Analysis

As in a water-base mud, the alkalinity, chloride content, and calcium content are important values to know. The reasons for needing to know them, however, are not the same because these chemicals are not contaminants, but produce desirable properties in an oil mud. The alkalinity value tells how much excess lime, or calcium hydroxide, the mud contains. Excess lime helps stabilize the emulsion and also neutralizes corrosive gases such as carbon dioxide and hydrogen sulfide.

The chloride test measures the amount of salt in the water phase, which controls shale sloughing. This value also is needed to correct data in solids content testing.

Calcium testing reveals the amount of calcium chloride, calcium hydroxide (lime), and anhydrite (drilled gypsum) in the oil mud.

Contaminants

Ordinary contaminants of water-base muds such as cement, salt, or anhydrite are not easily soluble in oil muds, so they have little or no detrimental effect on them. Solids, however, can be a problem. When properly formulated, oil muds do not make mud—that is, hydrate and disperse drilled solids. For this reason, preventing recirculation and regrinding of solids into the system is very important—the excess solids thicken an oil mud and lead to extra daily maintenance costs. Mechanical equipment such as shakers, mud cleaners, and centrifuges reduce solids buildup. API *Recommended Practice 13B-2* lists equations for calculating the lime, salt, and solids content of an oil mud from the data obtained in weight, chemical, and solids tests.

Other water-base mud contaminants such as carbon dioxide or hydrogen sulfide present no problem in oil muds that contain excess calcium hydroxide. Calcium hydroxide economically controls mud properties, including the oil-wetting properties that protect against corrosion from these gases. Often the operator chooses to use an oil mud solely for its effectiveness in the presence of these corrosives.

To summarize—

- Most tests on oil muds are the same as those for water-base muds
- Special considerations for oil mud testing: careful recording of mud temperature when testing funnel viscosity and testing for emulsion stability

---

▼
▼
▼

Treatment of Drilling Muds

▼
▼
▼

The testing that the crew and mud engineer do tells them if and when to treat the mud so that its properties are suitable to the drilling conditions. Changing drilling conditions can make it necessary to change the composition of the mud or change its weight.

Breakover

In many drilling operations, it becomes necessary to change the chemistry of the mud from one type to another. Such a change is referred to as a *conversion*, or *breakover*. The point in time when the properties of the mud actually change is called breakover. A breakover usually is quite a radical change in mud chemistry, and during breakover from one type of mud to another, there may be very severe "viscosity humps." For example, you can change a normal bentonite-based mud system to a salt-saturated system by adding salt. As you add the salt, the viscosity will increase really dramatically to very high levels, but eventually you reach a breakover point when the viscosity starts to decrease the more salt you add. At this point you are breaking over from one type of mud system to another.

Reasons for making a breakover include—

- to maintain a stable wellbore
- to provide a mud that will tolerate higher weight
- to drill salt formations
- to reduce the plugging of producing zones

In some cases it is dangerous to break over a mud in an open hole because of the high viscosities usually encountered. It is usually best to do a breakover in a cased hole before drilling the next hole section.

Weighting Up

The most important consideration when weighting up is to add the weighting material at a rate that will keep the mud weight constant in the suction tank while circulating. This is because the driller and the toolpusher need to know the exact weight of the mud going into the hole. Too heavy, and lost circulation can be a problem; too light, and a kick could occur. Careful density measurements of the mud in the suction pit will tell whether the weighting material is being added too slowly, too fast, or at the right rate.

A derrickhand in the field can figure how many sacks of barite to add and how fast to add them by approximate calculation. Or a mud engineer can make very accurate calculations. Which method is chosen depends on how critical the mix is.

Approximate Calculation (English measurements)

When the word comes down to weight up the mud, the derrickhand must calculate two things: (1) the amount of barite to add, and (2) how fast to add it.

Amount of Barite to Add per 100 Barrels

To calculate the amount of barite needed, first figure how many sacks of barite is needed for each 100 barrels of mud in the circulating system.

1. If the mud weighs less than 12.0 pounds per gallon (ppg), adding 60 sacks of barite will increase each 100 barrels (bbl) of mud by one pound per gallon.
2. If the mud weighs more than 12.0 ppg, divide the desired weight by 0.2 to find the number of sacks of barite needed to increase each 100 barrels of mud by one pound per gallon.

$$\frac{\text{desired weight}}{0.2} = \text{number of sacks per 100 barrels of mud}$$

Volume of Mud to be Treated

3. Next you need to know how many barrels of mud are in the system. Total volume of mud in the system includes the volume of mud in the hole plus the volume of mud in the tanks.
   a. You can calculate the volume of mud in the hole approximately or accurately. For approximate calculation, first calculate the approximate volume of the hole per 1,000 feet of depth. Do this by squaring the diameter of the hole in inches.

$$\text{hole volume (bbl) per 1,000 ft depth} = \text{diameter of hole (in)}^2$$

Next, multiply the hole volume per 1,000 feet by the depth in thousands of feet to find the total volume of the hole.

$$\text{total hole volume (bbl)} = \text{hole volume per 1,000 ft (bbl)} \times \text{depth of hole (1,000 ft)}$$

b. To use the accurate method, square the diameter of the hole.

$$\text{hole volume (bbl) per 1,000 ft depth} = \text{diameter of hole (in)}^2$$

Divide the result by 1,029 and multiply that number by the depth of the hole in feet.

$$\text{total hole volume (bbl)} = \frac{\text{hole volume per 1,000 ft (bbl)} \times \text{depth of hole (ft)}}{1{,}029}$$

c. Calculate the approximate volume of each tank by multiplying each tank's length by its width by the depth of the mud. (The depth of the mud remains constant all the time.)

$$\text{volume of tank (ft}^3\text{)} = \text{length (ft)} \times \text{width (ft)} \times \text{mud depth (ft)}$$

d. Add the volumes of the tanks together.

$$\text{total tank volume (ft}^3\text{)} = \text{volume of tank 1 (ft}^3\text{)} + \text{volume of tank 2 (ft}^3\text{), etc.}$$

e. Convert the volume in cubic feet to barrels so that the units of volume in the tanks match the units of volume in the hole. Each cubic foot of mud equals about 5.6 barrels, so divide by 5.6.

$$\text{volume of tanks (bbl)} = \frac{\text{volume of tanks (ft}^3\text{)}}{5.6}$$

f. Finally, add total hole volume to the total tank volume to find the volume of mud in the circulating system.

$$\text{volume of mud (bbl)} = \text{hole volume (bbl)} + \text{tank volume (bbl)}$$

4. Now multiply the results of step 2 by the results of step 3f and divide by 100 to find how much barite to add.

$$\text{total barite (sacks)} = \frac{\text{barite per 100 barrels mud (sacks)} \times \text{volume of mud (bbl)}}{100}$$

Rate of Addition

To figure how fast to add the barite, you need to calculate the *cycle time*, the time needed for the mud to make one complete circulating cycle from the suction intake at the pump to the bottom of the hole and back to pump suction. Cycle time is important because it is desirable to add all the barite during one circulating cycle.

5. First calculate the pump output in barrels per minute using one of these equations:

   pump output (bbl/min) = pump output (gal/min) × 0.024

   or

   pump output (bbl/min) = bbl/stroke × strokes/min

6. Calculate the cycle time by dividing the results of step 3f by the results of step 5.

$$\text{cycle time (min)} = \frac{\text{total volume of mud (bbl)}}{\text{pump output (bbl/min)}}$$

7. Figure out how many sacks to add per minute by dividing the number of sacks needed (step 4) by the cycle time (step 6).

$$\text{sacks per minute} = \frac{\text{total number of sacks}}{\text{cycle time (min)}}$$

Most tank hoppers can mix 5 to 10 sacks per minute without too much difficulty. To add barite as fast as 15 sacks per minute, for example, requires good hoppers and bulk barite storage tanks. If bulk barite is not available on the job, mud mixing must proceed at a slower rate. The usual procedure is to circulate at a slower rate and raise the mud weight over two or more circulation cycles.

Sample Problem

Let's solve a problem using the steps above. You want to raise the mud weight from 18.0 to 19.0 ppg, and you know the following about the system:

- The hole has a diameter of 9⅞ inches and is 10,000 feet deep.
- The system has two mud tanks. Each measures 8 feet by 20 feet and the mud is 7 feet deep.
- The pump output is 417 gallons per minute.

1. Because the mud weight is more than 12 ppg, begin with step 2.

2. To raise the weight from 18.0 to 19.0 ppg is to raise it 1.0 ppg. So calculate the number of sacks of barite per 100 barrels of mud.

$$\frac{19.0 \text{ ppg}}{0.2} = 95 \text{ sacks}$$

3. a. For approximate calculation, round off the hole diameter of 9⅞ inches and call it 10 inches.

   $$\text{hole volume per 1,000 ft depth} = 10^2 = 100 \text{ barrels}$$

   The hole is 10,000 feet deep, so multiply this number by 10.

   $$\text{total hole volume} = 100 \text{ bbl} \times 10{,}000 \text{ ft} = 1{,}000 \text{ bbl}$$

   b. For accurate calculation, do not round off the hole diameter.

   $$\text{hole volume per 1,000 ft depth} = 9.875^2 = 97.516 \text{ bbl}$$

   Multiply by the depth in feet and divide by 1,029

   $$\text{total hole volume} = \frac{97.516 \text{ bbl} \times 10{,}000 \text{ ft}}{1{,}029} = 948 \text{ bbl}$$

   c. Calculate the volume of each tank.

   $$\text{volume of each tank} = 8 \times 10 \times 7 = 560 \text{ ft}^3$$

   d. Add the volume of both tanks together.

   $$\text{volume of tanks} = 1{,}120 + 1{,}120 = 2{,}240 \text{ ft}^3$$

   e. Convert cubic feet to barrels.

   $$\text{volume of tanks} = \frac{2{,}240 \text{ ft}^3}{5.6} = 400 \text{ bbl}$$

   f. Calculate the total volume of mud in the system by adding the results of steps 3a and 3e.

   $$\text{total volume of mud} = 400 + 1{,}000 = 1{,}400 \text{ bbl}$$

4. Calculate the total number of sacks of barite to add.

   $$\text{total barite} = \frac{95 \times 1{,}400}{100} = 1{,}330 \text{ sacks}$$

   Now you know to add 1,330 sacks of barite to raise the weight of the mud to 19.0 ppg. Next figure how fast to add it.

5. Convert the pump output in gallons per minute to barrels per minute using the equation in step 5.

   $$\text{pump output} = 417 \text{ gal/min} \times 0.024 = 10 \text{ bbl/min}$$

6. Now calculate the cycle time using the results of steps 3f and 5.

$$\text{cycle time} = \frac{1{,}400 \text{ bbl}}{10 \text{ bbl/min}} = 140 \text{ minutes}$$

You will need to add all the barite over a period of 140 minutes.

7. Figure out how many sacks to add per minute by dividing the number of sacks needed (step 4) by the cycle time (step 6).

$$\text{sacks per minute} = \frac{1{,}330 \text{ sacks}}{140 \text{ min}} = 9.5 \text{ sacks per minute}$$

Approximate Calculation (SI Measurements)

When the word comes down to weight up the mud, the derrickhand must calculate two things: (1) the amount of barite to add, and (2) how fast to add it.

Amount of Weighting Material to Add

The calculations are different for barite and for clays because their specific gravities are different.

To calculate the amount of barite needed, first figure how many kilograms (kg) of barite are needed for each cubic metre (m^3) of mud in the circulating system.

1. Use the following equation when adding barite:

$$\begin{array}{c}\text{amount of barite} \\ \text{per m}^3 \text{ of mud}\end{array} = \frac{4{,}250 \text{ [desired density in kg/m}^3 - \text{initial density (kg/m}^3)]}{4{,}250 - \text{desired density in kg/m}^3}$$

2. Use this equation when adding clay:

$$\begin{array}{c}\text{amount of barite} \\ \text{per m}^3 \text{ of mud}\end{array} = \frac{2{,}500 \text{ [desired density in kg/m}^3 - \text{initial density (kg/m}^3)]}{2{,}500 - \text{desired density in kg/m}^3}$$

Volume of Mud to be Treated

3. Next you need to know how many cubic metres of mud are in the system. Total volume of mud in the system includes the volume of mud in the hole plus the volume of mud in the mud tanks.

$$\text{total volume} = \text{hole volume} + \text{tank volume}$$

a. Calculate the approximate volume of the hole by squaring the diameter of the hole and multiplying by the depth. Because the hole diameter will be in millimetres, convert first to metres by dividing by 1,000.

$$\text{hole diameter (m)} = \frac{\text{hole diameter (mm)}}{1{,}000}$$

$$\text{hole volume (m}^3\text{)} = \text{diameter of hole (m)}^2 \times \text{depth (m)}$$

b. Calculate the approximate volume of each tank by multiplying each tank's length by its width by the depth of the mud. (The depth of the mud remains constant all the time.)

$$\text{volume of tank (m}^3\text{)} = \text{length (m)} \times \text{width (m)} \times \text{mud depth (m)}$$

c. Add the volumes of all the active tanks together.

$$\text{total tank volume} = \text{volume of tank 1} + \text{volume of tank 2, etc.}$$

d. Finally, add total hole volume to the total tank volume to find the volume of mud in the circulating system.

$$\text{volume of mud (m}^3\text{)} = \text{hole volume (m}^3\text{)} + \text{tank volume (m}^3\text{)}$$

4. Now multiply the result of step 2 by the result of step 3d to figure how much barite or clay to add.

$$\text{total weighting material (kg)} = \text{weighting material per m}^3 \text{ mud (kg)} \times \text{volume of mud (m}^3\text{)}$$

5. Convert kilograms to sacks. If a sack contains 40 kilograms of weighting material, divide the result of step 4 by 40.

$$\text{sacks of weighting material} = \frac{\text{kilograms of weighting material}}{\text{40 kg/sack}}$$

Rate of Addition

To figure how fast to add the barite, you need to calculate the *cycle time*, the time needed for the mud to make one complete circulating cycle from the suction intake at the pump to the bottom of the hole and back to pump suction. Cycle time is important because it is desirable to add all the barite during one circulating cycle.

6. First calculate the pump output in cubic metres per minute using one of these equations:

   $$\text{pump output (m}^3\text{/min)} = \text{pump output (litres/min)} \times .001$$

   or

   $$\text{pump output (m}^3\text{/min)} = \text{litres/stroke} \times \text{strokes/min} \times .001$$

7. Calculate the cycle time by dividing the results of step 3d by the results of step 5.

   $$\text{cycle time (min)} = \frac{\text{total volume of mud (m}^3\text{)}}{\text{pump output (m}^3\text{/min)}}$$

8. Figure out how many sacks to add per minute by dividing the number of sacks needed (step 5) by the cycle time (step 7).

   $$\text{sacks per minute} = \frac{\text{total number of sacks}}{\text{cycle time (min)}}$$

Sample Problem

Let's solve a problem using the steps above. You want to raise the mud weight from 1,138.3 to 1,234.1 kilograms per cubic metre, and you know the following about the system:

- The hole has a diameter of 250 millimetres and is 3,000 metres deep.
- The system has two mud tanks. Each measures 2.5 metres by 6 metres and the mud is 2 metres deep.
- The pump output is 1,580 litres per minute.
- The weighting material is barite.

1. Because the weighting material is barite, use the equation in step 1.

   $$\text{amount of barite per m}^3\text{ of mud} = \frac{4{,}250\ (1{,}234.1\ \text{kg/m}^3 - 1{,}138.3\ \text{kg/m}^3)}{4{,}250 - 1{,}234.1\ \text{kg/m}^3} = 135\ \text{kg/}$$

2. a. Calculate the approximate volume of the hole after converting its diameter from millimetres to metres.

   $$\text{hole diameter} = \frac{250\ \text{mm}}{1{,}000} = 0.25\ \text{m}$$

   $$\text{hole volume} = (0.25\text{m})^2 \times 3{,}000\ \text{m} = 750\ \text{m}^3$$

   b. Calculate the volume of each tank.

   $$\text{volume of tank} = 2.5\ \text{m} \times 6\ \text{m} \times 2\ \text{m} = 30\ \text{m}^3$$

c. Add the volumes of the two tanks.

$$\text{total tank volume} = 30\ m^3 + 30\ m^3 = 60\ m^3$$

d. Finally, add total hole volume to the total tank volume to find the volume of mud in the circulating system.

$$\text{volume of mud} = 750\ m^3 + 60\ m^3 = 810\ m^3$$

3. Now multiply the result of step 1 by the result of step 2d to figure how much barite to add.

$$\text{total barite} = 135\ kg/m^3 \times 830\ m^3 = 112{,}050\ kg$$

4. Convert kilograms to sacks.

$$\text{sacks of weighting material} = \frac{112{,}050\ kg}{40\ kg/sack} = 2{,}801\ \text{sacks}$$

Now you know to add 2,801 sacks of barite to raise the weight of the mud to 1,234.1 kilograms per cubic metre. Next figure how fast to add it.

5. First calculate the pump output in cubic metres per minute using one of these equations:

$$\text{pump output } (m^3/min) = 1{,}580\ (\text{litres/min}) \times .001 = 1.58\ m^3$$

6. Calculate the cycle time by dividing the result of step 2d by the result of step 5.

$$\text{cycle time (min)} = \frac{830\ m^3}{\text{pump output } (m^3/min)}$$

7. Figure out how many sacks to add per minute by dividing the number of sacks needed (step 4) by the cycle time (step 6).

$$\text{sacks per minute} = \frac{2{,}801\ \text{sacks}}{\text{cycle time (min)}}$$

Calculating Increased Volume

Adding weighting material also increases the volume of mud, which the derrickhand needs to take into consideration. The total mud volume must not exceed the holding capacity of the system.

English Calculation

The addition of each 100 sacks of barite increases mud volume by 6.7 barrels. So to calculate how much the total volume will increase, divide the number of sacks to be added by 6.7.

$$\text{volume increase (bbl)} = \frac{\text{number of sacks of barite added}}{6.7}$$

Using the sample problem above, how much will the total mud volume increase?

$$\text{volume increase} = \frac{1{,}330 \text{ sacks}}{6.7} = 198.5 \text{ bbl}$$

SI Calculation

To calculate the volume increase, divide the amount of weighting material added by the constant for the type of material. For barite, the constant is 4,250, and for clay it is 2,500.

$$\text{volume increase (m}^3\text{)} = \frac{\text{kg barite added}}{4{,}250}$$

$$\text{volume increase (m}^3\text{)} = \frac{\text{kg clay added}}{2{,}500}$$

Using the sample problem above, how much will the total mud volume increase?

$$\text{volume increase} = \frac{1{,}743 \text{ kg barite}}{4{,}250} = 0.41 \text{ m}^3$$

Using Tables

Table 13 shows the number of 100-pound sacks of barite required to raise the weight of 100 barrels of mud (of a given initial weight in ppg) to a desired weight (in ppg). Table 14 shows an equivalent metric table. These are useful for estimating how much barite to add to weight up a mud a certain amount, but derrickhands do not generally have them on hand at the rig.

For example, Table 13 shows that raising 13.0-ppg mud to 15.0 pounds per gallon requires 147 sacks of barite per 100 barrels of mud. Similarly, Table 14 shows that raising density from 1,300 to 1,500 kilograms per cubic metre requires about 300 kilograms of barite per cubic metre of mud.

Table 13
Number of 100-lb Sacks of Barite per 100 bbl of Mud
for Desired Mud Weight in ppg

Desired Weight (ppg)	100-lb sacks barite per ppg-increase per 100 bbl mud
9.5	57.2
10.0	58.3
10.5	59.5
11.0	60.7
11.5	61.9
12.0	63.2
12.5	64.6
13.0	65.9
13.5	67.4
14.0	69.0
14.5	70.6
15.0	72.3
15.5	74.1
16.0	76.0
16.5	77.9
17.0	80.0
17.5	82.2
18.0	84.5

Table 14
Number of 45.4-kg Sacks of Barite per 10 m^3 of Mud for Desired Mud Weight in kg/m^3

Desired Weight (kg/m^3)	**45-kg sacks barite per kg/m^3-increase per 10 m^3 mud**
1,140	35.75
1,200	36.4
1,260	37.2
1,320	37.9
1,380	38.7
1,440	39.5
1,500	40.4
1,560	41.2
1,620	42.1
1,680	43.1
1,740	44.1
1,800	45.4
1,860	46.3
1,920	47.5
1,980	48.7
2,040	50.0
2,100	51.4
2,160	52.8

Calculating Hydrostatic Pressure

A related calculation that the derrickhand makes is to determine hydrostatic pressure downhole. To do this, measure the weight of the mud with a mud balance, then multiply the weight by the depth of the fluid column (depth of the hole). Depending on whether the balance measures mud weight in pounds per gallon (ppg), pounds per cubic foot (pcf or lb/ft^3), or kilograms per metre (kg/m), choose one of the following equations to calculate the hydrostatic pressure.

$$\text{hydrostatic pressure (psi)} = \text{depth (ft)} \times \text{mud weight (ppg)} \times 0.052$$

$$\text{hydrostatic pressure (psi)} = \text{depth (ft)} \times \text{mud weight (pcf)} \times 0.00695$$

$$\text{hydrostatic pressure (kPa)} = \text{depth (m)} \times \text{mud weight (kg/m)}$$

Water-Back

Occasionally, the operator wants to reduce the mud weight rather than increase it. For example, a heavier mud may be needed to drill a high-pressure formation. Then, after casing is set, the high-pressure formation is sealed off behind the casing. If the formations below the casing have normal pressure, the operator may then wish to decrease the mud weight.

In water-base muds, adding water is the usual way to reduce mud weight. This is one meaning of the term *water-back*. (Water-back also refers to adding water in order to decrease the solids content of the mud.) Tables 3 and 4 show the effect of adding water on mud weight.

Calculating Water-Back (English Measurements)

The following formula approximates the volume of water needed to reduce mud weight:

$$x = \frac{V(W_1 - W_2)}{W_2 - 8.34}$$

where

x = barrels of water needed
V = original volume of mud, in barrels
W_1 = initial mud weight, in pounds per gallon
W_2 = desired mud weight, in pounds per gallon.

For example, if the total volume of mud in the system is 1,000 bbl and you want to reduce the mud weight from 12.0 ppg to 11.0 ppg, then calculate as follows:

$$x = \frac{1{,}000\ (12.0 - 11.0)}{11.0 - 8.34} = 376 \text{ bbl of water}$$

Calculating Water-Back (SI Measurements)

The following formula approximates the volume of water needed to reduce mud weight:

$$x = \frac{V(D_1 - D_2)}{D_2 - 1{,}000}$$

where

x = cubic metres of water needed
V = original volume of mud, in cubic metres
D_1 = initial mud weight, in kilograms per litre
D_2 = desired mud weight, in kilograms per litre.

For example, if the total volume of mud in the system is 143 cubic metres and you want to reduce the mud weight from 1,320 kg/l to 1,080 kg/l, then calculate as follows:

$$x = \frac{143\ (1{,}320 - 1{,}080)}{1{,}080 - 1{,}000} = 429 \text{ cubic metres of water}$$

These formulas do not consider the effect of solids settling due to decreased viscosity. When the crew adds large amounts of water, they must also add materials to increase the viscosity to prevent the weighting material from settling out of the mud. They add them slowly through the hopper to avoid plugging the hopper.

To summarize—

- Breakover is the point when mud properties actually change while altering the mud chemistry
- The most important consideration when weighting up is to add the weighting material at a rate that will keep the mud weight constant in the suction tank while circulating
  - Adding weighting material increases the volume of mud in the system
- Water-back refers to adding water in order to decrease viscosity or the solids content of the mud

---

Appendix A

Calculations Used in Mud Work

From *IADC Drilling Manual*. Used with permission.

A. Volume of Mud in System

One of the first concerns of the well employee is the volume of mud in the system. This includes the mud in the pit and the mud in the hole. Another fact useful to know is the amount of reserve mud available in storage tanks or pits.

bbl of mud in system = bbl of mud in pit + bbl of mud in hole

$$\text{vol. of mud in pit (bbl)} = \frac{\text{length (ft)} \times \text{width (ft)} \times \text{depth (ft)}}{5.6\ \text{ft}^3/\text{bbl}}$$

B. Capacity of Hole

The most frequently used formula for estimating the volume of mud in the hole is—

$$\text{vol. (bbl/1,000 ft of hole)} = (\text{diameter of hole in inches})^2$$

C. Mud Cycling Time

In mud conditioning, it is frequently important to know the time required for the mud to make a cycle from the pump suction to the bottom of the hole and back to the pump suction. It is almost always desirable to add weight material or chemicals at such a rate that the mud will make at least one complete cycle during the treatment. Two factors are involved in cycling time during the treatment: (1) the volume of mud in the hole and the active surface volume, and (2) the output rate of the slush pump.

Cycle Time

pump output (gpm × 0.024) = pump output (bbl/min)

pump output (bbl/min) = bbl/stroke × strokes/min

$$\text{time of complete circulation} = \frac{\text{bbl mud in hole + active surface vol. (bbl)}}{\text{pump output (bbl/min)}}$$

D. Hydrostatic Pressure Equivalents

hydrostatic pressure (psi) = mud weight (ppg) × 0.052 × depth (ft)

hydrostatic pressure (psi) = mud weight (lb/ft³) × 0.00695 × depth (ft)

avg. formation fluid press. (psi) (fresh water gradient) = 0.433 × depth (ft)

avg. formation fluid press. (psi) (saltwater gradient) = 0.465 × depth (ft)

mud weight (ppg) × 7.4805 = mud weight (lb/ft³)

mud weight (lb/ft³) × 0.2337 = mud weight (ppg)

mud weight (ppg) ÷ 8.33 = specific gravity of mud

E. Quantities of Mud Materials

1. Increase in weight

Barite: sp gr 4.25

35.5 ppg or 1,490 lb/bbl

Approximate: In the weight range 9.0–12.0 ppg, 60 sacks barite increases weight of 110 bbl mud 1 ppg.

For weights above 12.0 ppg, divide the desired or final weight by 0.2 to find number of sacks barite to raise weight of 100 bbl mud 1 ppg.

Sacks barite needed to increase weight of 100 bbl mud:

$$x = \frac{1{,}490\,(W_2 - W_1)}{35.5 - W_2}$$

where

W_1 = present mud weight (ppg)
W_2 = desired mud weight (ppg).

2. Increase in volume from addition of weight material

Approximate: 15 sacks barite = 1 bbl of volume, or 100 sacks barite = 6.64 bbl of volume. Barrels increase in volume in weighting up 100 bbl of mud is as follows:

$$V = \frac{100\,(W_2 - W_1)}{35.5 - W_2}$$

where

V = increase in volume (bbl)
W_1 = initial mud weight (ppg)
W_2 = final mud weight (ppg).

3. To determine quantity of water necessary to obtain a given weight reduction (neglecting settling effect)

$$x = \frac{(W_1 - W_2)\,V_1}{W_2 - 8.33}$$

where

x = bbl of water to be added
W_1 = original mud weight (ppg)
W_2 = desired mud weight (ppg)
V_1 = original mud volume (bbl).

F. Annular or Rising Velocity of Mud

The upward velocity of mud in the annulus between the drill pipe and the wall of the hole is an important consideration for the drilling fluid function of removing cuttings. The usual expression of this velocity is in feet per minute.

annular volume (bbl/ft) = capacity open hole (bbl/ft) – capacity of drill pipe and displacement of drill pipe (bbl/ft)

$$\text{annular velocity (ft/min)} = \frac{\text{pump output (bbl/min)}}{\text{annular vol. (bbl/ft)}}$$

$$\text{annular velocity (ft/min)} = \frac{\text{gal/min} \times 0.024}{\text{annular vol. (bbl/ft)}}$$

Appendix B

Basic Mud Chemicals

Viscosifiers
- Bentonite
- Attapulgite
- Polymers

Viscosity-Reducing Chemicals
- Lignosulfonate
- Lignites
- Tannates
- Phosphates

Fluid-Loss Reducers
- Starches
- CMC
- Bentonite
- Synthetic polymers
- Lignites
- Lignosulfonate

Weighting Materials
- Barite
- Galena
- Calcium carbonate
- Dissolved salts
- Iron oxide

Swelling Inhibitors
- Salt
- Encapsulating agent
- Lime
- Gypsum

Emulsifiers
- Lignites
- Lignosulfonate
- Detergents

Lost-Circulation Materials
- Granular
- Fibrous
- Flaked
- Slurries

Special Additives
- Flocculants
- Corrosion control
- Defoamer
- pH control
- Mud lubricant
- Antidifferential sticking material

Glossary

▼
▼
▼

A

abnormal pressure *n*: formation pressure exceeding or falling below the pressure to be expected at a given depth. Normal pressure increases approximately 0.465 pounds per square inch per foot of depth or 10.5 kilopascals per metre of depth. Thus, normal pressure at 1,000 feet is 465 pounds per square inch; at 1,000 metres it is 10,500 kilopascals. See *pressure gradient.*

acid gas *n*: a gas that forms an acid when mixed with water. In petroleum production and processing, the most common acid gases are hydrogen sulfide and carbon dioxide. Both cause corrosion, and hydrogen sulfide is very poisonous.

acidity *n*: the quality of being acid. Relative acid strength of a liquid is measured by pH. A liquid with a pH below 7 is acid. See *pH.*

active mud tank *n*: one of usually two, three, or more mud tanks that holds drilling mud that is being circulated into a borehole during drilling. They are called active tanks because they hold mud that is currently being circulated.

additive *n*: 1. in general, a substance added in small amounts to a larger amount of another substance to change some characteristic of the latter. In the oil industry, additives are used in lubricating oil, fuel, drilling mud, and cement. 2. in cementing, a substance added to cement to change its characteristics to satisfy specific conditions in the well. A cement additive may work as an accelerator, retarder, dispersant, or other reactant.

adhesion *n*: a force of attraction that causes molecules of one substance to cling to those of a different substance.

aerated mud *n*: drilling mud into which air or gas is injected. Aeration with air or gas reduces the density of the mud and allows for faster drilling rates. The lighter aerated mud does not develop as much pressure on bottom as a normal mud. The lower pressure allows the cuttings made by the bit to easily break away from the bit's cutters; the cutters therefore always contact fresh, undrilled formation.

aggregate *n*: a group of two or more particles held together by strong forces. Aggregates are stable with normal stirring, shaking, or handling; they may be broken by treatment such as ball milling a powder or shearing a suspension.

aging test *n*: a procedure whereby a product may be subjected to intensified but controlled conditions of heat, pressure, radiation, or other variables to produce, in a short time, the effects of long-time storage or use under normal conditions.

agitator *n*: a motor-driven paddle or blade used to mix the liquids and solids in drilling mud.

air drilling *n*: a method of rotary drilling that uses compressed air as the circulation medium. The conventional method of removing cuttings from the wellbore is to use a flow of water or drilling mud. Compressed air removes the cuttings with equal or greater efficiency. The rate of penetration is usually increased considerably when air drilling is used; however, a principal problem in air drilling is the penetration of formations containing water, since the entry of water into the system reduces the ability of the air to remove the cuttings.

alkali *n*: a substance having marked basic (alkaline) properties, such as a hydroxide of an alkali metal. See *base*.

alkalinity *n*: the quality of being basic. The strength of a liquid's alkalinity is measured by pH; a pH above 7 is alkaline. See *pH*.

American Petroleum Institute (API) *n*: oil trade organization (founded in 1920) that is the leading standardizing organization for oilfield drilling and producing equipment. It maintains departments of transportation, refining, marketing, and production in Washington, D.C. It offers publications regarding standards, recommended practices, and bulletins. Address: 1220 L St., NW; Washington, D.C. 20005; (202) 682-8000.

anhydrite *n*: the common name for anhydrous calcium sulfate, $CaSO_4$.

annular pressure *n*: fluid pressure in an annular space, as around tubing within casing.

annular space *n*: the space between two concentric circles. In the petroleum industry, it is usually the space surrounding a pipe in the wellbore; sometimes termed the annulus.

annular velocity *n*: the rate at which mud is traveling in the annular space of a drilling well.

annulus *n*: see *annular space*.

antidifferential sticking additive *n*: a chemical added to the drilling fluid to minimize the possibility of the drill stem becoming stuck to the side of the hole.

API *abbr*: American Petroleum Institute.

API gravity *n*: the measure of the density or gravity of liquid petroleum products in the United States; derived from relative density. API gravity is expressed in degrees, 10° API being equivalent to 1.0, the specific gravity of water. See *specific gravity*.

apparent viscosity *n*: the viscosity of a drilling fluid as measured with a direct-indicating, or rotational, viscometer.

asbestos *n*: a fibrous mineral used to build viscosity in saltwater drilling fluids. It is carcinogenic when breathed, and closely regulated in the U.S.

attapulgite *n*: a fibrous clay mineral that is a viscosity-building substance; used principally in saltwater-base drilling muds. Also called fuller's earth.

B

back-pressure *n*: the pressure maintained on equipment or systems through which a fluid flows.

bactericide *n*: a substance that kills bacteria.

ball up *v*: to collect a mass of sticky consolidated material, usually drill cuttings, on drill pipe, drill collars, bits, and so forth. A bit with such material attached to it is called a balled-up bit. Balling up is frequently the result of inadequate pump pressure or insufficient drilling fluid.

barite *n*: barium sulfate, a mineral frequently used to increase the weight or density of drilling mud. Its specific gravity is 4.2 (i.e., it is 4.2 times denser than water). See *barium sulfate, mud.*

barium sulfate *n*: a chemical compound of barium, sulfur, and oxygen ($BaSO_4$), which may form a tenacious scale that is very difficult to remove. Also called barite.

barrel (bbl) *n*: 1. a measure of volume for petroleum products in the United States. One barrel is the equivalent of 42 U.S. gallons or 0.15899 cubic metres (9,702 cubic inches). One cubic metre equals 6.2897 barrels. 2. the cylindrical part of a sucker rod pump in which the pistonlike plunger moves up and down. Operating as a piston inside a cylinder, the plunger and barrel create pressure energy to lift well fluids to the surface.

baryte *n*: variation of barite. See *barite.*

base *n*: a substance capable of reacting with an acid to form a salt. A typical base is sodium hydroxide (caustic), with the chemical formula NaOH. For example, sodium hydroxide combines with hydrochloric acid to form sodium chloride (a salt) and water. This reaction is written chemically as $NaOH + HCl \rightarrow NaCl + H_2O$.

basicity *n*: pH value above 7 and the ability to neutralize or accept protons from acids.

bbl *abbr*: barrel.

bentonite *n*: a colloidal clay, composed primarily of montmorillonite, that swells when wet. Because of its gel-forming properties, bentonite is a major component of water-base drilling muds. See *gel, mud.*

bentonite extenders *n pl*: a group of polymers that can maintain or increase the viscosity of bentonite while flocculating other clay solids in the mud. With bentonite extenders, desired viscosity can often be maintained using only half the amount of bentonite that would otherwise be required.

BHP *abbr*: bottomhole pressure.

BHT *abbr*: bottomhole temperature.

biopolymer *n*: a polymer produced by the action of a particular strain of bacteria on carbohydrates, used for increasing apparent viscosity and yield point with moderately good filtration control. Also called X-C polymer.

biopolymer mud *n*: a drilling mud formulated with a biopolymer.

bit *n*: the cutting or boring element used in drilling oil and gas wells. The bit consists of a cutting element and a circulating element. The cutting element is steel teeth, tungsten carbide buttons, industrial diamonds, or polycrystalline diamonds (PDCs). The circulating element permits the passage of drilling fluid and utilizes the hydraulic force of the fluid stream to improve drilling rates. In rotary drilling, several drill collars are joined to the bottom end of the drill pipe column, and the bit is attached to the end of the drill collars.

blind drilling *n*: a drilling operation in which the drilling fluid is not returned to the surface; rather, it flows into an underground formation. Sometimes blind-drilling techniques are resorted to when lost circulation occurs.

blooey line *n*: the discharge pipe from a well being drilled by air drilling. The blooey line is used to conduct the air or gas used for circulation away from the rig to reduce the fire hazard as well as to transport the cuttings a suitable distance from the well. See *air drilling*.

blowout *n*: an uncontrolled flow of gas, oil, or other well fluids into the atmosphere or into an underground formation. A blowout, or gusher, can occur when formation pressure exceeds the pressure applied to it by the column of drilling fluid. See *kick*.

blowout preventer *n*: one of several valves installed at the wellhead to prevent the escape of pressure either in the annular space between the casing and the drill pipe or in open hole (i.e., hole with no drill pipe) during drilling or completion operations. Blowout preventers on land rigs are located beneath the rig at the land's surface; on jackup or platform rigs, at the water's surface; and on floating offshore rigs, on the seafloor.

BOP *abbr*: blowout preventer.

bottomhole *n*: the lowest or deepest part of a well. *adj*: pertaining to the bottom of the wellbore.

bottomhole pressure *n*: the pressure at the bottom of a borehole. It is caused by the hydrostatic pressure of the wellbore fluid and, sometimes, by any back-pressure held at the surface, as when the well is shut in with blowout preventers. When mud is being circulated, bottomhole pressure is the hydrostatic pressure plus the remaining circulating pressure required to move the mud up the annulus.

bottomhole temperature *n*: temperature measured in a well at a depth at the midpoint of the thickness of the producing zone.

break circulation *v*: to start the mud pump for restoring circulation of the mud column. Because the stagnant drilling fluid has thickened or gelled during the period of no circulation, high pump pressure is usually required to break circulation.

breakover *n*: the change in the chemistry of a mud from one type to another. Also called conversion.

brine *n*: water that has a large quantity of salt, especially sodium chloride, dissolved in it; salt water.

C

C *abbr*: Celsius (formerly centigrade). See *Celsius scale*.

cake *n*: see *filter cake*.

calcium contamination *n*: dissolved calcium ions in a drilling fluid in sufficient concentration to impart undesirable properties, such as flocculation, reduction in yield of bentonite, and increased fluid loss. See also *calcium hydroxide*, *calcium sulfate*, *gypsum*, *lime*.

calcium hydroxide *n*: the active ingredient of slaked (hydrated) lime, and the main constituent in cement (when wet). Referred to as "lime" in field terminology. Its symbol is $Ca(OH)_2$.

calcium sulfate *n*: a chemical compound of calcium, sulfur, and oxygen, $CaSO_4$. Although sometimes considered a contaminant of drilling fluids, it may at times be added to them to produce certain properties. Like calcium carbonate it forms scales in water-handling facilities, which may be hard to remove. See *anhydrite, gypsum*.

calcium-treated mud *n*: a freshwater drilling mud using calcium oxide (lime) or calcium sulfate (gypsum) to retard the hydrating qualities of shale and clay formations, thus facilitating drilling. Calcium-treated muds resist salt and anhydrite contamination but may require further treatment to prevent gelation (solidification) under the high temperatures of deep wells.

calibration *n*: the process of checking a measuring instrument to determine its deviation from a standard measurement.

carboxymethyl cellulose *n*: a nonfermenting cellulose product used in drilling fluids to combat calcium contamination and to lower the fluid loss of the mud.

caustic *n*: see *caustic soda*.

caustic soda *n*: sodium hydroxide, NaOH. It is used to maintain an alkaline pH in drilling mud and in petroleum fractions. Also called caustic.

caving *n*: collapsing of the walls of the wellbore. Also called sloughing.

cavings *n pl*: particles that fall off (are sloughed from) the wall of the wellbore. Compare *cuttings*.

Celsius scale *n*: the metric scale of temperature measurement used universally by scientists. On this scale, 0° represents the freezing point of water and 100° its boiling point at a barometric pressure of 760 mm. The Celsius scale was formerly called the centigrade scale; now, however, the term "Celsius" is preferred in the International System of Units (SI).

centigrade scale *n*: see *Celsius scale*.

centimetre (cm) *n*: a unit of length in the metric system equal to one-hundredth of a metre (10^{-2} metre).

centipoise *n*: a unit of viscosity.

chemical barrel *n*: a container in which various chemicals are mixed prior to addition to drilling fluid.

chemical protective clothing *n*: clothing that is designed to protect against a specific chemical hazard (i.e., suits or aprons made of or coated with chemical-resistant materials like butyl rubber, neoprene, or polyvinyl chloride).

chemicals *n pl*: in drilling-fluid terminology, a chemical is any material that produces changes in the viscosity, yield point, gel strength, fluid loss, and surface tension.

chemical treatment *n*: any of many processes in the oil industry that involve the use of a chemical to effect an operation. Some chemical treatments are acidizing, crude oil demulsification, corrosion inhibition, paraffin removal, scale removal, drilling fluid control, refinery and plant processes, cleaning and plugging operations, chemical flooding, and water purification.

circulate *v*: to pass from one point throughout a system and back to the starting point. For example, drilling fluid is circulated out of the suction pit, down the drill pipe and drill collars, out the bit, up the annulus, and back to the pits while drilling proceeds.

circulating density *n*: see *equivalent circulating density.*

circulating fluid *n*: see *drilling fluid, mud.*

circulating pressure *n*: the pressure generated by the mud pumps and exerted on the drill stem.

circulating rate *n*: the volume flow rate of the circulating drilling fluid usually expressed in gallons or barrels per minute in the United States. Elsewhere, it is expressed in cubic metres per minute.

circulation *n*: the movement of drilling fluid out of the mud pits, down the drill stem, up the annulus, and back to the mud pits. See *normal circulation, reverse circulation.*

clay *n*: a group of hydrous aluminum silicate minerals (clay minerals).

clay extender *n*: any of several substances—usually organic compounds of high molecular weight—that, when added in low concentrations to a bentonite or to certain other clay slurries, will increase the viscosity of the system. See *low-solids mud.*

clay yield *n*: the number of barrels of a liquid slurry of a given viscosity that can be made from a ton of clay. Clays are often classified as either high- or low-yield. A ton of high-yield clay yields more slurry of a given viscosity than a low-yield clay.

Clean Water Act (CWA) *n*: a U.S. law that regulates the discharge of toxic and nontoxic pollutants into the surface waters of the United States. Under the jurisdiction of both the EPA and the Army Corps of Engineers, CWA's short-term goal is to make surface waters safe for recreation, fishing, and other uses. Its long-term goal is to eliminate *all* harmful discharges into surface waters.

clear brine *n*: a drilling fluid made up mainly of chemical salts, such as sodium chloride, calcium chloride, or potassium chloride. Clear brine contains little or no clay or other solid material and is virtually transparent. It is often used when drilling into a producing formation because clear brine minimizes formation damage. See *formation damage.*

clear water drilling *n*: drilling operations in which plain water (usually salt water) is used as the circulating fluid.

closed system *n*: in circulation, a system in which no drilling fluid is discarded to a reserve pit. Drilling companies may use closed systems when environmental regulations do not permit any contaminants to be released. Companies are also discovering that closed systems can be economical when using expensive drilling muds.

cm *abbr*: centimetre.

cm^2 *abbr*: square centimetre.

cm^3 *abbr*: cubic centimetre.

CMC *abbr*: carboxymethyl cellulose.

colloid *n*: 1. a substance whose particles are so fine that they will not settle out of suspension or solution and cannot be seen under an ordinary microscope. 2. the mixture of a colloid and the liquid, gaseous, or solid medium in which it is dispersed.

colloidal *adj*: pertaining to a colloid, i.e., involving particles so minute (less than 2 microns) that they are not visible through optical microscopes. Bentonite is an example of a colloidal clay.

colloidal phase *n*: the part of a drilling mud that consists of microscopic particles that can react with the liquid. The main colloidal solids in most drilling muds are clays. Also called the reactive phase.

colorimetric testing *n*: method of testing pH using chemically treated paper strips that change color.

condition *v*: to treat drilling mud with additives to give it certain properties. Sometimes the term applies to water used in boilers, drilling operations, and so on. To condition and circulate mud is to ensure that additives are distributed evenly throughout a system by circulating the mud while it is being conditioned. See *mud conditioning*.

connate water *n*: water retained in the pore spaces, or interstices, of a formation from the time the formation was created. Also called interstitial water.

consolidated shales *n pl*: compacted beds of clays and silts that are firm.

contaminant *n*: a material, usually a mud component, that becomes mixed with cement slurry during displacement and that affects it adversely.

continuous phase *n*: the liquid phase of a drilling mud. It may be either water or oil.

conversion *n*: the change in the chemistry of a mud from one type to another. Reasons for making a conversion may be (1) to maintain a stable wellbore, (2) to provide a mud that will tolerate higher weight, or density, (3) to drill soluble formations, or (4) to protect producing zones. Also called a breakover.

corrosion *n*: any of a variety of complex chemical or electrochemical processes, e.g., rust, by which metal is destroyed through reaction with its environment.

corrosion control *n*: the measures used to prevent or reduce the effects of corrosion. These practices can range from simply painting metal, to isolating it from moisture and chemicals and insulating it from galvanic currents, to cathodic protection, in which a galvanic or impressed direct electric current renders a pipeline cathodic, thus causing it to be a negative element in the circuit. The use of chemical inhibitors and closed systems are other examples of corrosion control.

corrosion control agent *n*: a chemical added to drilling fluid to minimize corrosion to the drill stem.

corrosion inhibitor *n*: a chemical substance that minimizes or prevents corrosion in metal equipment.

corrosion-resisting steel *n*: a steel alloy that contains chromium and nickel and that does not corrode as quickly as normal steel alloys.

corrosive *adj*: a corrosion agent, e.g., acid.

crew *n*: 1. the workers on a drilling or workover rig, including the driller, the derrickhand, and the rotary helpers. 2. any group of oilfield workers.

crosslinking *n*: a process of molecular bridging of polymers with other chemical substances that alters viscosity and shear rates to enhance lifting of bit cuttings and increase drilling rates.

cubic centimetre (cm^3) *n*: a commonly used unit of volume measurement in the metric system equal to 10^{-6} cubic metre, or 1 millilitre. The volume of a cube whose edge is 1 centimetre.

cubic metre (m^3) *n*: a unit of volume measurement in the metric system, replacing the previous standard unit known as the barrel, which was equivalent to 35 imperial gallons or 42 U.S. gallons. The cubic metre equals approximately 6.2898 barrels.

cuttings *n pl*: the fragments of rock dislodged by the bit and brought to the surface in the drilling mud. Washed and dried cuttings samples are analyzed by geologists to obtain information about the formations drilled.

D

deflocculation *n*: the dispersion of solids that have stuck together in drilling fluid, usually by means of chemical thinners. See *flocculation.*

defoamer *n*: any chemical that prevents or lessens frothing or foaming in another agent.

degasser *n*: the device used to remove unwanted gas from a liquid, especially from drilling fluid.

density *n*: the mass or weight of a substance per unit volume. For instance, the density of a drilling mud may be 10 pounds per gallon, 74.8 pounds/cubic foot, or 1,198.2 kilograms/cubic metre. Specific gravity, relative density, and API gravity are other units of density. See *mud weight.*

derrickhand *n*: the crew member who handles the upper end of the drill string as it is being hoisted out of or lowered into the hole. He or she is also responsible for the circulating machinery and the conditioning of the drilling fluid.

desander *n*: a centrifugal device for removing sand from drilling fluid to prevent abrasion of the pumps. It may be operated mechanically or by a fast-moving stream of fluid inside a special cone-shaped vessel, in which case it is sometimes called a hydrocyclone. Compare *desilter.*

desilter *n*: a centrifugal device for removing very fine particles, or silt, from drilling fluid to keep the amount of solids in the fluid at the lowest possible point. Usually, the lower the solids content of mud, the faster is the rate of penetration. The desilter works on the same principle as a desander. Compare *desander.*

differential pressure *n*: the difference between two fluid pressures; for example, the difference between the pressure in a reservoir and in a wellbore drilled in the reservoir.

differential sticking *n*: a condition in which the drill stem becomes stuck against the wall of the wellbore because part of the drill stem (usually the drill collars) has become embedded in the filter cake. Necessary conditions for differential sticking are a permeable formation and a pressure differential across a nearly impermeable filter cake and drill stem. Also called wall sticking.

direct-indicating viscometer *n*: a rotational device powered by means of an electric motor or handcrank. Used to determine the apparent viscosity, plastic viscosity, yield point, and gel strengths of drilling fluids. Commonly called a "V-G meter."

directional drilling *n*: intentional deviation of a wellbore from the vertical. Although wellbores are normally drilled vertically, it is sometimes necessary or advantageous to drill at an angle from the vertical. Controlled directional drilling makes it possible to reach subsurface areas laterally remote from the point where the bit enters the earth. It often involves the use of deflection tools.

dispersant *n*: a substance added to cement that chemically wets the cement particles in the slurry, allowing the slurry to flow easily without much water.

dispersed phase *n*: that part of a drilling mud—clay, shale, barite, and other solids—that is dispersed throughout a liquid or gaseous medium, forming the mud.

disperser *n*: see *emulsifying agent.*

dispersion *n*: a suspension of extremely fine particles in a liquid (such as colloids in a colloidal solution).

downhole *adj, adv*: pertaining to the wellbore.

downstream *adv, adj*: in the direction of flow in a stream of fluid moving in a line.

drilled solids *n pl*: the fine particles in drilling mud drilled by the bit.

drill-in fluid *n*: a drilling fluid specially formulated to minimize formation damage as the borehole penetrates the producing zone. See *formation damage.*

drilling blind *n*: drilling without mud returns, as in the case of severe lost circulation.

drilling break *n*: a sudden increase in the drill bit's rate of penetration. It sometimes indicates that the bit has penetrated a high-pressure zone and thus warns of the possibility of a kick.

drilling fluid *n*: a liquid, air, or natural gas that is circulated through the wellbore during rotary drilling operations.

drilling mud *n*: a liquid drilling fluid containing additives to alter its properties. See *drilling fluid, mud.*

drilling rate *n*: the speed with which the bit drills the formation; usually called the rate of penetration (ROP).

dusting *n*: drilling with dry air or gas.

E

ECD *abbr*: equivalent circulating density.

electrical stability test *n*: a test for oil muds that determines the stability of the emulsion.

emulsifier *n*: a material that causes water and oil to form an emulsion. Water normally occurs separately from oil; if, however, an emulsifying agent is present, the water becomes dispersed in the oil as tiny droplets. Or, rarely, the oil may be dispersed in the water. In either case, the emulsion must be treated to separate the water and the oil. Also called surfactant, wetting agent.

emulsifying agent *n*: a material that causes water and oil to form an emulsion. Water normally occurs separately from oil; if, however, an emulsifying agent is present, the water becomes dispersed in the oil as tiny droplets. Or, rarely, the oil may be dispersed in the water. In either case, the emulsion must be treated to separate the water and the oil.

emulsion *n*: a mixture in which one liquid, termed the dispersed phase, is uniformly distributed (usually as minute globules) in another liquid, called the continuous phase or dispersion medium. In an oil-in-water emulsion, the oil is the dispersed phase and the water the dispersion medium; in a water-in-oil emulsion, the reverse holds.

emulsion mud *n*: see *oil-in-water emulsion mud.*

endpoint *n*: the point marking the end of one stage of a process. In filtrate analysis, the endpoint is the point at which a particular result is achieved through titration.

entrained gas *n*: formation gas that enters the drilling fluid in the annulus. See *gas-cut mud.*

environment *n*: 1. the sum of the physical, chemical, and biological factors that surround an organism. 2. the water, air, and land and the interrelationship that exists among and between water, air, and land and all living things. 3. as defined by the U.S. government, the navigable waters, the waters of the contiguous zone, the ocean waters, and any other surface water, groundwater, drinking water supply, land surface, subsurface strata, or ambient air within the United States.

Environmental Protection Agency (EPA) *n*: a U.S. agency that was created in 1970 from a variety of existing agencies. The EPA administers air pollution, water pollution, pesticide, solid waste, noise control, drinking water, and toxic substances acts. It also has major research responsibilities. Address: 401 M Street SW; Washington, D.C. 20460; (202) 260-2090.

equivalent circulating density (ECD) *n*: the increase in bottomhole pressure expressed as an increase in pressure that occurs only when mud is being circulated. Because of friction in the annulus as the mud is pumped, bottomhole pressure is slightly, but significantly, higher than when the mud is not being pumped. ECD is calculated by dividing the annular pressure loss by 0.052, dividing that by true vertical depth, and adding the result to the mud weight. Also called circulating density, mud-weight equivalent.

ES *abbr*: electrical stability.

extended bentonites *n pl*: selected bentonites treated with chemical polymers. See *extender.*

extender *n*: 1. a substance added to drilling mud to increase viscosity without adding clay or other thickening material. 2. an additive that assists in getting greater yield from a sack of cement. The extender acts by requiring more water than required by neat cement.

F

F *abbr*: Fahrenheit. See *Fahrenheit scale.*

Fahrenheit scale *n*: a temperature scale devised by Gabriel Fahrenheit, in which 32° represents the freezing point and 212° the boiling point of water at standard sea-level pressure.

Fann V-G™ meter *n*: trade name of a device used to record and measure at different speeds the flow properties of plastic fluids (such as the viscosity and gel strength of drilling fluids).

filter *n*: a porous medium through which a fluid is passed to separate particles of suspended solids from it.

filter cake *n*: 1. compacted solid or semisolid material remaining on a filter after pressure filtration of mud with a standard filter press. Thickness of the cake is reported in thirty-seconds of an inch or in millimetres. 2. the layer of concentrated solids from the drilling mud or cement slurry that forms on the walls of the borehole opposite permeable formations; also called wall cake or mud cake.

filter cake thickness *n*: a measurement of the solids deposited on filter paper in thirty-seconds of an inch during a standard 30-minute API filter test. In certain areas the filter cake thickness is a measurement of the solids deposited on filter paper for 7.5 minutes.

filter press *n*: a device used to test the filtration properties of drilling mud. See *filtration qualities*.

filtrate *n*: 1. a fluid that has been passed through a filter. 2. the liquid portion of drilling mud that is forced into porous and permeable formations next to the borehole.

filtration *n*: the process of filtering a fluid. Oil workers sometimes use filtration to mean fluid loss.

filtration loss *n*: the escape of the liquid part of a drilling mud into permeable formations. Also called fluid loss.

filtration qualities *n pl*: the filtration characteristics of a drilling mud. In general, these qualities are inverse to the thickness of the filter cake deposited on the face of a porous medium and the amount of filtrate allowed to escape from the drilling fluid into or through the medium.

flocculant *n*: see *flocculating agent*.

flocculating agent *n*: material or chemical agent that enhances flocculation. Also called flocculant.

flocculation *n*: the coagulation of solids in a drilling fluid, produced by special additives or by contaminants.

flocs *abbr*: flocculates.

flow properties *n pl*: the properties of a drilling mud that have to do with its flow—viscosity, gel strength, and yield point.

flow rate *n*: the speed, or velocity, of fluid flow through a pipe or vessel.

flow sensor *n*: a tool inserted into a pipeline or other container that can sense the flow of fluid within the container.

fluid *n*: a substance that flows and yields to any force tending to change its shape. Liquids and gases are fluids.

fluid loss *n*: the unwanted migration of the liquid part of the drilling mud or cement slurry into a formation, often minimized or prevented by the blending of additives with the mud or cement.

fluid-loss additive *n*: a compound added to cement slurry or drilling mud to prevent or minimize fluid loss.

foam *n*: a two-phase system, similar to an emulsion, in which the dispersed phase is a gas or air.

foam drilling *n*: see *mist drilling*.

foaming agent *n*: a chemical used to lighten the water column in gas wells, in oilwells producing gas, and in drilling wells in which air or gas is used as the drilling fluid so that the water can be forced out with the air or gas to prevent its impeding the production or drilling rate. See *mist drilling*.

formation *n*: a bed or deposit composed throughout of substantially the same kind of rock; often a lithologic unit. Each formation is given a name, frequently as a result of the study of the formation outcrop at the surface and sometimes based on fossils found in the formation.

formation breakdown *n*: the fracturing of a formation from excessive borehole pressure.

formation breakdown pressure *n*: the pressure at which a formation will fracture.

formation competency *n*: the ability of the formation to withstand applied pressure. Also called formation integrity.

formation damage *n*: the reduction of permeability in a reservoir rock caused by the invasion of drilling fluid and treating fluids to the section adjacent to the wellbore. It is often called skin damage.

formation pressure *n*: the force exerted by fluids in a formation, recorded in the hole at the level of the formation with the well shut in. Also called reservoir pressure or shut-in bottomhole pressure.

formation strength *n*: the ability of a formation to resist fracture from pressures created by fluids in a borehole.

formation water *n*: 1. the water originally in place in a formation. 2. any water that resides in the pore spaces of a formation.

fracture *n*: a crack or crevice in a formation, either natural or induced.

fracture gradient *n*: the pressure gradient (psi/foot) at which a formation accepts whole fluid from the wellbore. Also called frac gradient.

fracture pressure *n*: the pressure at which a formation will break down, or fracture.

fracture zone *n*: zone of naturally occurring fissures or fractures that can pose problems with lost circulation.

fresh water *n*: 1. water that has little or no salt dissolved in it. 2. underground water, generally located near the surface, that does not contain a large amount of salt and from which most underground drinking water supplies are drawn. 3. inland surface water, such as lakes, streams, and ponds, that is not salty.

friction loss *n*: a reduction in the pressure of a fluid caused by its motion against an enclosed surface (such as a pipe). As the fluid moves through the pipe, friction between the fluid and the pipe wall and within the fluid itself creates a pressure loss. The faster the fluid moves, the greater are the losses.

ft *abbr*: foot.

ft^2 *abbr*: square foot.

ft^3 *abbr*: cubic foot.

ft^3/bbl *abbr*: cubic feet per barrel.

ft^3/min *abbr*: cubic feet per minute.

funnel viscosity *n*: viscosity as measured by the Marsh funnel, based on the number of seconds it takes for 61 cubic inches (1,000 cubic centimetres) of drilling fluid to flow through the funnel.

G

gal *abbr*: gallon.

galena (PbS) *n*: lead sulfide. Technical grades (specific gravity about 7) are used for increasing the density of drilling fluids to points impractical or impossible with barite.

gallon *n*: a unit of measure of liquid capacity that equals 3.785 litres and has a volume of 231 cubic inches (0.00379 cubic metres). A gallon of water weighs 8.34 pounds (3.8 kilograms) at 60°F (16°C). The imperial gallon, formerly used in Great Britain, equals approximately 1.2 U.S. gallons.

gal/min *abbr*: gallons per minute.

gas *n*: a compressible fluid that completely fills any container in which it is confined.

gas buster *n*: a mud-gas separator.

gas-cut mud *n*: a drilling mud that contains entrained formation gas, giving the mud a characteristically fluffy texture. When entrained gas is not released before the fluid returns to the well, the weight or density of the fluid column is reduced. Because a large amount of gas in mud lowers its density, gas-cut mud must be treated to reduce the chance of a kick.

gas cutting *n*: a process in which gas becomes entrained in a liquid.

gas drilling *n*: a method of drilling that uses natural gas as the drilling fluid. See *air drilling*.

gel *n*: 1. a semisolid, jellylike state assumed by some colloidal dispersions at rest. When agitated, the gel converts to a fluid state. 2. a nickname for bentonite. *v*: to take the form of a gel; to set.

gel strength *n*: a measure of the ability of a colloidal dispersion to develop and retain a gel form, based on its resistance to shear. The gel, or shear, strength of a drilling mud determines its ability to hold solids in suspension. Sometimes bentonite and other colloidal clays are added to drilling fluid to increase its gel strength.

geopressure *n*: abnormally high pressure exerted by some subsurface formations. The deeper the formation, the higher the pressure it exerts on a wellbore drilled into it.

geopressured shales *n pl*: impermeable shales, highly compressed by overburden pressure, that are characterized by large amounts of formation fluids and abnormally high pore pressure.

geothermal gradient *n*: the increase in the temperature of the earth with increasing depth. It averages about 1°F per 60 feet (1°C per 18.3 metres), but may be considerably higher or lower.

gpm *abbr*: gallons per minute, when referring to rate of flow.

gravity *n*: see *API gravity, relative density, specific gravity.*

groundwater *n*: water that seeps through soil and fills pores of underground rock formations; the source of water in springs and wells.

gumbo *n*: any relatively sticky formation (such as clay) encountered in drilling.

gyp *n*: (slang) gypsum.

gyp mud *n*: drilling mud that is treated with gypsum to provide a source of soluble calcium in the filtrate to obtain desirable mud properties for drilling in shale or clay formations.

gypsum *n*: a naturally occurring crystalline form of calcium sulfate in which each molecule of calcium sulfate is combined with two molecules of water. See *anhydrite, calcium sulfate.*

H

H_2S *form*: hydrogen sulfide.

hard water *n*: water that contains dissolved calcium and magnesium salts.

hazardous *adj*: involving or exposing one to risk. The lists of material or waste that are considered hazardous vary from agency to agency and from regulation to regulation: hazardous materials in transport are regulated by DOT; hazardous substances in the workplace are regulated by OSHA; hazardous waste is regulated under EPA's RCRA; toxic substances are regulated under EPA's TSCA; hazardous air pollutants are regulated under EPA's CAA; and so on. The hazardous list for that regulation is tailored to that purpose.

head *n*: 1. the height of a column of liquid required to produce a specific pressure. See *hydraulic head.* 2. for centrifugal pumps, the velocity of flowing fluid converted into pressure expressed in feet or metres of flowing fluid. Also called velocity head. 3. that part of a machine (such as a pump or an engine) that is on the end of the cylinder opposite the crankshaft. Also called cylinder head.

high-pH mud *n*: a drilling fluid with a pH range above 10.5, i.e., a high-alkalinity mud.

high-yield drilling clay *n*: a classification given to a group of commercial drilling-clay preparations having a yield of 35 to 50 barrels (5,565 to 7,950 litres) per ton and intermediate between bentonite and low-yield clays. Usually prepared by peptizing low-yield calcium montmorillonite clays or, in a few cases, by blending some bentonite with the peptized low-yield clay.

hydrate *n*: a hydrocarbon and water compound that is formed under reduced temperature and pressure in gathering, compression, and transmission facilities for gas. Hydrates often accumulate in troublesome amounts and impede fluid flow. They resemble snow or ice. *v*: to enlarge by taking water on or in.

hydraulic head *n*: the force exerted by a column of liquid expressed by the height of the liquid above the point at which the pressure is measured. Although "head" refers to distance or height, it is used to express pressure, since the force of the liquid column is directly proportional to its height. Also called head or hydrostatic head. Compare *hydrostatic pressure.*

hydraulics *n*: 1. the branch of science that deals with practical applications of water or other liquid in motion. 2. the planning and operation of a rig hydraulics program, coordinating the power of circulating fluid at the bit with other aspects of the drilling program so that bottomhole cleaning is maximized.

hydrophilic *adj*: having a strong tendency to bind or absorb water, which results in swelling and formation of a reversible gel.

hydrostatic head *n*: see *hydrostatic pressure.*

hydrostatic pressure *n*: the force exerted by a body of fluid at rest. It increases directly with the density and the depth of the fluid and is expressed in pounds per square inch or kilopascals. The hydrostatic pressure of fresh water is 0.433 pounds per square inch per foot (9.792 kilopascals per metre) of depth. In drilling, the term refers to the pressure exerted by the drilling fluid in the wellbore. In a water drive field, the term refers to the pressure that may furnish the primary energy for production.

hydroxide *n*: a designation that is given for basic compounds containing the hydroxide (OH) radical. When these substances are dissolved in water, they increase the pH of the solution. See *base.*

I

incompetent formation *n*: a formation composed of materials that are not bound together. It may produce sand along with hydrocarbons if preventive measures are not taken, or it may slough or cave around the bit or drill stem when a hole is drilled into it.

initial gel strength *n*: the maximum reading (deflection) taken from a direct-reading viscometer after the fluid has been quiescent for 10 seconds. It is reported in pounds/100 square feet. See *gel strength.*

invert-emulsion mud *n*: an oil mud in which fresh or salt water is the dispersed phase and diesel, crude, or some other oil is the continuous phase. Also called invert-oil mud. See *oil mud.*

invert-oil mud *n*: see *invert-emulsion mud.*

K

kick *n*: an entry of water, gas, oil, or other formation fluid into the wellbore during drilling. It occurs because the pressure exerted by the column of drilling fluid is not great enough to overcome the pressure exerted by the fluids in the formation drilled. If prompt action is not taken to control the kick, or kill the well, a blowout may occur.

kick fluids *n pl*: oil, gas, water, or any combination that enters the borehole from a permeable formation.

kill *v*: 1. in drilling, to control a kick by taking suitable preventive measures (e.g., to shut in the well with the blowout preventers, circulate the kick out, and increase the weight of the drilling mud). 2. in production, to stop a well from producing oil and gas so that reconditioning of the well can proceed. Production is stopped by circulating a kill fluid into the hole.

kill fluid *n*: drilling mud of a weight great enough to equal or exceed the pressure exerted by formation fluids.

kill mud *n*: see *kill fluid.*

kilograms per cubic centimetre (kg/m^3) *n*: a measure of the density, or weight, of a fluid (such as drilling mud).

kilopascal (kPa) *n*: 1,000 pascals. The SI metric unit of measurement for pressure and stress and a component in the measurement of viscosity. A pascal is equal to a force of 1 newton acting on an area of 1 square metre.

kPa *abbr*: kilopascal.

L

lb *abbr*: pound.

lb/bbl *abbr*: pounds per barrel.

lb/ft^3 *abbr*: pounds per cubic foot.

LCM *abbr*: lost circulation material.

lignins *n pl*: naturally occurring special lignites, e.g., leonardite, that are produced by strip mining from special lignite deposits. Used primarily as thinners and emulsifiers.

lignosulfonate *n*: an organic drilling fluid additive derived from by-products of a paper-making process using sulfite. It minimizes fluid loss and reduces mud viscosity.

lime *n*: a caustic solid that consists primarily of calcium oxide (CaO). Many forms of CaO are called lime, including the various chemical and physical forms of quicklime, hydrated lime, and even calcium carbonate. Limestone is sometimes called lime.

lime hydrate *n*: an alkaline chemical also known as lime, slaked lime, or calcium hydroxide, $Ca(OH)_2$.

lime mud *n*: a drilling mud that is treated with lime to provide a source of soluble calcium in the filtrate to obtain desirable mud properties for drilling in shale or clay formations.

liquid *n*: a state of matter in which the shape of the given mass depends on the containing vessel, but the volume of the mass is independent of the vessel. A liquid is a fluid that is almost incompressible.

liquid phase *n*: in drilling fluids, that part of the fluid that is liquid. Normally, the liquid phase of a drilling fluid is water, oil, or a combination of water and oil.

lost circulation *n*: the quantities of whole mud lost to a formation, usually in cavernous, fissured, or coarsely permeable beds. Evidenced by the complete or partial failure of the mud to return to the surface as it is being circulated in the hole. Lost circulation can lead to a blowout and, in general, can reduce the efficiency of the drilling operation. Also called lost returns.

lost circulation additives *n pl*: materials added to the mud in varying amounts to control or prevent lost circulation. Classified as fiber, flake, or granular.

lost circulation material (LCM) *n*: a substance added to cement slurries or drilling mud to prevent the loss of cement or mud to the formation. See *lost circulation additives*.

lost returns *n pl*: see *lost circulation*.

low clay-solids mud *n*: heavily weighted muds whose high solids content (a result of the large amounts of barite added) necessitates the reduction of clay solids.

low-solids mud *n*: a drilling mud that contains a minimum amount of solid material and that is used in rotary drilling when possible because it can provide fast drilling rates.

low-yield clay *n*: commercial clay chiefly of the calcium montmorillonite type and having a yield of approximately 15 barrels per ton (2,385 litres per tonne).

lubricant *n*: a substance—usually petroleum-based—that is used to reduce friction between two moving parts.

M

m *abbr*: metre.

m^2 *abbr*: square metre.

m^3 *abbr*: cubic metre.

make mud *v*: the tendency of soft formations penetrated by a wellbore to mix with clear water being circulated into the well.

Marsh funnel *n*: a calibrated funnel used in field tests to determine the viscosity of drilling mud.

methylene blue *n*: a dye that colors the reactive clays (bentonite and/or drilled solids) in a drilling mud sample for determining the percentage of clay in the sample.

min *abbr*: minute.

mist drilling *n*: a drilling technique that uses air or gas to which a foaming agent has been added.

mixing mud *n*: preparation of drilling fluids from a mixture of water and other fluids and one or more of the various dry mud-making materials such as clay and chemicals.

mixing tank *n*: any tank or vessel used to mix components of a substance (as in the mixing of additives with drilling mud).

mix mud *v*: to prepare drilling fluids from a mixture of water or other liquids and any one or more of the various dry mud-making materials (such as clay, weighting materials, and chemicals).

mm *abbr*: millimetre.

mm^2 *abbr*: square millimetre.

mm^3 *abbr*: cubic millimetre.

montmorillonite *n*: a clay mineral often used as an additive to drilling mud. It is a hydrous aluminum silicate capable of reacting with such substances as magnesium and calcium. See *bentonite*.

mud *n*: the liquid circulated through the wellbore during rotary drilling and workover operations. Although it was originally a suspension of earth solids (especially clays) in water, the mud used in modern drilling operations is a more complex, three-phase mixture of liquids, reactive solids, and nonreactive solids. The liquid phase may be fresh water, diesel oil, or crude oil and may contain one or more conditioners. See *drilling fluid, drilling mud.*

mud additive *n*: any material added to drilling fluid to change some of its characteristics or properties.

mud analysis *n*: examination and testing of drilling mud to determine its physical and chemical properties.

mud balance *n*: a beam balance consisting of a cup and a graduated arm carrying a sliding weight and resting on a fulcrum. It is used to determine the density or weight of drilling mud.

mud cake *n*: the sheath of mud solids that forms on the wall of the hole when liquid from mud filters into the formation. Also called filter cake or wall cake.

mud column *n*: the borehole when it is filled or partially filled with drilling mud.

mud conditioning *n*: the treatment and control of drilling mud to ensure that it has the correct properties. Conditioning may include the use of additives, the removal of sand or other solids, the removal of gas, the addition of water, and other measures to prepare the mud for conditions encountered in a specific well.

mud density *n*: see *mud weight.*

mud density recorder *n*: a device that automatically records the weight or density of drilling fluid as it is being circulated in a well.

mud engineer *n*: an employee of a drilling fluid supply company whose duty it is to test and maintain the drilling mud properties that are specified by the operator.

mud-flow indicator *n*: a device that continually measures and may record the flow rate of mud returning from the annulus and flowing out of the mud return line. If the mud does not flow at a fairly constant rate, a kick or lost circulation may have occurred.

mud-gas separator *n*: a device that removes entrained gas from drilling mud.

mud inhibitor *n*: a substance, such as salt, potassium chloride, or calcium sulfate, added to drilling mud to minimize the hydration (swelling) of formations with which the mud is in contact.

mud line *n*: 1. in offshore operations, the seafloor. 2. a mud return line.

mud-making *n*: the ability of a formation to mix with clear water being circulated in a wellbore and create a natural drilling mud.

mud man *n*: see *mud engineer.*

mud misting *n*: a type of mist drilling where a thin mud slurry instead of clear water carries the foamer.

mud pit *n*: originally, an open pit dug in the ground to hold drilling fluid or waste materials discarded after the treatment of drilling mud. Now it is another name for a mud tank. See *mud tank.*

mud program *n*: a plan or procedure, with respect to depth, for the type and properties of drilling fluid to be used in drilling a well. Some factors that influence the mud program are the casing program and such formation characteristics as type, competence, solubility, temperature, and pressure.

mud return line *n*: a trough or pipe that is placed between the surface connections at the wellbore and the shale shaker and through which drilling mud flows on its return to the surface from the hole. Also called flow line.

mud solids *n pl*: the solid components of drilling mud. They may be added intentionally (barite, for example), or they may be introduced into the mud from the formation as the bit drills ahead. The term is usually used to refer to the latter.

mud still *n*: instrument used to distill oil, water, and other volatile materials in a mud to determine oil, water, and total solids contents in volume-percent. Also called a retort.

mud system *n*: the composition and characteristics of the drilling mud used on a particular well.

mud tank *n*: a steel tank to hold drilling fluid or waste materials discarded after the treatment of drilling mud. For some drilling operations, mud pits are used for suction to the mud pumps, settling of mud sediments, and storage of reserve mud. They may still be referred to as mud pits. Also called slush tank.

mud up *v*: to add solid materials (such as bentonite or other clay) to a drilling fluid composed mainly of clear water to obtain certain desirable properties.

mud weight *n*: a measure of the density of a drilling fluid expressed as pounds per gallon, pounds per cubic foot, or kilograms per cubic metre. Mud weight is directly related to the amount of pressure the column of drilling mud exerts at the bottom of the hole.

mud-weight equivalent *n*: see *equivalent circulating density.*

N

natural clays *n pl*: clays that are encountered when drilling various formations; they may or may not be incorporated purposely into the mud system.

natural gas *n*: a highly compressible, highly expansible mixture of hydrocarbons with a low specific gravity and occurring naturally in a gaseous form. Besides hydrocarbon gases, natural gas may contain appreciable quantities of nitrogen, helium, carbon dioxide, hydrogen sulfide, and water vapor. Although gaseous at normal temperatures and pressures, the gases making up the mixture that is natural gas are variable in form and may be found either as gases or as liquids under suitable conditions of temperature and pressure.

natural mud *n*: a drilling fluid containing essentially clay and water; no special or expensive chemicals or conditioners are added. Also called conventional mud.

nonreactive phase *n*: the solids in drilling mud that do not react with the liquid.

normal circulation *n*: the smooth, uninterrupted circulation of drilling fluid down the drill stem, out the bit, up the annular space between the pipe and the hole, and back to the surface. Compare *reverse circulation.*

normal formation pressure *n*: formation fluid pressure equivalent to about 0.465 pounds per square inch per foot (10.5 kilopascals per metre) of depth from the surface. If the formation pressure is 4,650 pounds per square inch (32,062 kilopascals) at 10,000 feet (3,048 metres), it is considered normal.

normal pore pressure *n*: pore pressure where the pressure of any water in the rock equals the hydrostatic pressure of a column of salt water of the same depth.

normal pressure gradient *n*: the pressure developed by a column of fluid as the depth of the column increases when the column contains a fluid of normal density. This gradient varies from area to area, but along the Gulf Coast of the United States, it is considered to be 0.465–0.468 psi/foot (10.53–10.59 kPa/metre), which is the pressure developed by the salt water that naturally occurs in the formations of this area. See *normal formation pressure*.

O

Occupational Safety and Health Administration (OSHA) *n*: a U.S. government agency that conducts research into the causes of occupational diseases and accidents. It is responsible for administration of the certification of respiratory safety equipment. Address: Department of Labor; 200 Constitution Avenue, NW; Washington, D.C. 20210; (202) 523-1452.

oil-base mud *n*: a drilling or workover fluid in which oil is the continuous phase and which contains from less than 2 percent and up to 5 percent water. This water is spread out, or dispersed, in the oil as small droplets. See *invert-emulsion mud, oil mud*.

oil-in-water emulsion mud *n*: a water-base mud in which water is the continuous phase and oil is the dispersed phase. The oil is spread out, or dispersed, in the water in small droplets, which are tightly emulsified so that they do not settle out. Because of its lubricating abilities, an oil-emulsion mud increases the drilling rate and ensures better hole conditions than other muds. Compare *oil mud*.

oil mud *n*: a drilling mud, e.g., oil-base mud and invert-emulsion mud, in which oil is the continuous phase. It is useful in drilling certain formations that may be difficult or costly to drill with water-base mud. Compare *oil-in-water emulsion mud*.

operator *n*: the person or company, either proprietor or lessee, actually operating an oilwell or lease, generally the oil company that engages the drilling contractor.

organophilic clay *n*: a clay treated so that it is not hydrophilic for use in oil muds. Also called organic clay, amine clay.

overbalance *n*: the extent to which the hydrostatic pressure of the mud column exceeds formation pressure.

overbalanced drilling *n*: drilling in which the hydrostatic pressure of the mud column exceeds formation pressure.

overburden *n*: the strata of rock that overlie the stratum of interest in drilling.

overburden pressure *n*: the pressure exerted by the rock strata on a formation of interest. It is usually considered to be about 1 pound per square inch per foot (22.621 kilopascals per metre).

P

pH *abbr*: an indicator of the acidity or alkalinity of a substance or solution, represented on a scale of 0–14, 0–6.9 being acidic, 7 being neither acidic nor basic (i.e., neutral), and 7.1–14 being basic. These values are based on hydrogen ion content and activity.

phase *n*: a portion of a physical system that is liquid, gas, or solid, that is homogeneous throughout, that has definite boundaries, and that can be separated from other phases. The three phases of drilling mud are the liquid or continuous phase, the reactive or colloidal phase, and the nonreactive phase.

plastic deformation *n*: a phenomenon in which a decrease in the diameter of the borehole occurs because a plastic substance, such as salt, is being drilled. Since the temperature of the salt is higher than normal for a given depth (high temperature causes the salt to expand) and since mud weight is usually decreased when drilling such formations to maintain high penetration rates (thus decreasing the hydrostatic pressure exerted on the salt), the section of borehole in the salt zone can expand and possibly stick the drill stem. Also called salt squeeze.

plastic flow *n*: see *plastic fluid.*

plastic fluid *n*: a complex, non-Newtonian fluid in which the shear force is not proportional to the shear rate. Most drilling muds are plastic fluids.

plasticity *n*: the ability of a substance to be deformed without rupturing.

plastic viscosity *n*: an absolute flow property indicating the flow resistance of certain types of fluids. It is a measure of shear stress.

polymer *n*: a substance that consists of large molecules formed from smaller molecules in repeating structural units (monomers). In oilfield operations, various types of polymers are used to thicken drilling mud, fracturing fluid, acid, water, and other liquids. See *polymer mud.*

polymer mud *n*: a drilling mud to which a polymer has been added to increase the viscosity of the mud.

pore pressure *n*: the force exerted by fluids in a formation, recorded in the hole at the level of the formation with the well shut in. Also called formation pressure, reservoir pressure, or shut-in bottomhole pressure.

pounds per cubic foot *n*: a measure of the density of a substance (such as drilling fluid).

pounds per gallon (ppg) *n*: a measure of the density, or weight, of a fluid (such as drilling mud).

pounds per 100 square feet *n*: a measure of force, for example, of the gel strength of drilling mud.

pounds per square inch (psi) *n*: a measure of pressure.

ppg *abbr*: pounds per gallon.

precipitate *n*: a substance, usually a solid, that separates from a fluid because of a chemical or physical change in the fluid. *v*: to separate in this manner.

precipitation *n*: the production of a separate liquid phase from a mixture of gases (e.g., rain), or of a separate solid phase from a liquid solution, as in the precipitation of calcite cement from water in the interstices of rock.

pressure *n*: the force that a fluid (liquid or gas) exerts uniformly in all directions within a vessel, pipe, hole in the ground, and so forth, such as that exerted against the inner wall of a tank or that exerted on the bottom of the wellbore by a fluid. Pressure is expressed in terms of force exerted per unit of area, as pounds per square inch, or in kilopascals.

pressure gradient *n*: a scale of pressure differences in which there is a uniform variation of pressure from point to point. For example, the pressure gradient of a column of water is about 0.433 psi/ft (9.794 kPa/m) of vertical elevation.

pressure loss *n*: the drilling fluid's loss of hydraulic pressure after it leaves the pump. Some pressure is lost due to friction, but the main loss occurs when the fluid leaves the bit nozzles. See *friction loss.*

pressure surge *n*: a sudden and usually of short-duration increase in pressure. When pipe or casing is run into a hole too rapidly, an increase in the hydrostatic pressure results, which may be great enough to create lost circulation.

psi *abbr*: pounds per square inch.

psi/ft *abbr*: pounds per square inch per foot.

Q

quebracho *n*: a South American tree that is a source of tannin extract, which was extensively used as a thinning agent for drilling mud, but is seldom used today.

quicklime *n*: calcium oxide, CaO, used in certain oil-base muds to neutralize the organic acid.

R

rate of penetration (ROP) *n*: the speed with which the bit drills the formation.

reactive phase *n*: the part of a drilling mud that consists of microscopic particles that can react with the liquid. The main reactive solids in most drilling muds are clays. Also called the colloidal phase.

reactive solids content *n*: the amount of water-absorbent material in the drilling fluid.

reactivity *n*: a measure of one substance's ability to chemically react with other substances.

red-lime mud *n*: a water-base clay mud containing caustic soda and tannates to which lime has been added. Also called red mud.

red mud *n*: see *red-lime mud.*

relative density *n*: 1. the ratio of the weight of a given volume of a substance at a given temperature to the weight of an equal volume of a standard substance at the same temperature. For example, if 1 cubic inch of water at 39°F (3.9°C) weighs 1 unit and 1 cubic inch of another solid or liquid at 39°F weighs 0.95 unit, then the relative density of the substance is 0.95. In determining the relative density of gases, the comparison is made with the standard of air or hydrogen. 2. the ratio of the mass of a given volume of a substance to the mass of a like volume of a standard substance, such as water or air.

relaxed invert-emulsion mud *n*: an invert-emulsion mud with a higher oil-to-water ratio than is usual.

representative sample *n*: a small portion extracted from the total volume of material that contains the same proportions of the various flowing constituents as the total volume of liquid being transferred. The precision of extraction must be equal to or better than the method used to analyze the sample.

retort *n*: an instrument used to distill oil, water, and other volatile materials in a mud to determine oil, water, and total solids contents in volume-percent. Also called a still.

returns *n pl*: the mud, cuttings, and so forth, that circulate up the hole to the surface.

reverse circulation *n*: the course of drilling fluid downward through the annulus and upward through the drill stem, in contrast to normal circulation in which the course is downward through the drill stem and upward through the annulus. Seldom used in open hole, but frequently used in workover operations. Also referred to as "circulating the short way," since returns from bottom can be obtained more quickly than in normal circulation. Compare *normal circulation.*

reverse emulsion *n*: a relatively rare oilfield emulsion composed of globules of oil dispersed in water. Most oilfield emulsions consist of water dispersed in oil.

rheology *n*: the study of the flow of gases and liquids of special importance to mud engineers and reservoir engineers.

ROP *abbr*: rate of penetration.

S

sack *n*: a container for cement, bentonite, ilmenite, barite, caustic, and so forth. Sacks (bags) contain the following amounts:

Cement	94 pounds (42.6 kilograms) (1 cubic foot)
Bentonite	100 pounds (45.5 kilograms)
Ilmenite	100 pounds
Barite	100 pounds

salinity *n*: a measure of the amount of salt dissolved in a liquid.

salt barrel *n*: a 55-gallon (208-litre) drum modified to salt-saturate the water going into circulation to prevent the dissolution of formation salt when building mud volume.

salt mud *n*: 1. a drilling mud in which the water has an appreciable amount of salt (usually sodium or calcium chloride) dissolved in it. Also called saltwater mud or saline drilling fluid. 2. a mud with a resistivity less than or equal to the formation water resistivity.

salt water *n*: a water that contains a large quantity of salt, i.e., brine.

saltwater clay *n*: see *attapulgite.*

saltwater mud *n*: see *salt mud.*

sample mud *n*: drilling fluid formulated so that it will not alter the properties of the cuttings the fluid carries up the well.

samples *n pl*: 1. the well cuttings obtained at designated footage intervals during drilling. From an examination of these cuttings, the geologist determines the type of rock and formations being drilled and estimates oil and gas content. 2. small quantities of well fluids obtained for analysis.

sampling *n*: 1. the taking of a representative sample of fluid from a tank or pipeline to measure its temperature, specific gravity, and S&W content. 2. the process of cutting a core or pieces of core for analysis.

sand *n*: in drilling, an abrasive material composed of small grains formed from the disintegration of any type of rock. Sand is larger than 74 microns (.074 mm or .003 inch) and smaller than 2,000 microns (2 millimetres, or 0.08 inch) in diameter.

saturated saltwater *n*: water that has all the salt dissolved in it that is possible at a given temperature.

screen set *n*: an instrument for measuring the sand content of drilling mud.

seawater mud *n*: a special class of saltwater muds in which sea water is used as the fluid phase.

seconds API *n*: the time in seconds that it takes 1 quart of drilling mud to flow out of a Marsh funnel. It is a measure of the mud's viscosity.

shale *n*: a fine-grained sedimentary rock composed mostly of consolidated clay or mud. Shale is the most frequently occurring sedimentary rock.

shear *n*: action or stress that results from applied forces and that causes or tends to cause two adjoining portions of a substance or body to slide relative to each other in a direction parallel to their plane of contact.

shear rate *n*: the rate of shear due to stress.

shear strength *n*: see *gel strength*.

shear stress *n*: force applied to a liquid to cause it to flow.

shear thinning *n*: a phenomenon in which the mud's viscosity decreases due to shear stress.

shut-in *adj*: shut off to prevent flow. Said of a well, plant, pump, and so forth, when valves are closed at both inlet and outlet.

SI *abbr*: International System of Units, or metric system.

silt *n*: a material that exhibits little or no swelling and whose particle size ranges from 2 to 74 microns. Dispersed clays and barite fall into this particle-size range.

slip velocity *n*: 1. the rate at which drilled solids tend to settle in the borehole as a well is being drilled. 2. difference between the annular velocity of the fluid and the rate at which a cutting is removed from the hole.

sloughing (pronounced "sluffing") *n*: collapsing of the walls of the wellbore. Also called caving.

sloughing hole *n*: a condition wherein shale that has absorbed water from the drilling fluid expands, sloughs off, and falls downhole. A sloughing hole can jam the drill string and block circulation.

slow-release inhibitor *n*: corrosion-preventive substance that is released into production fluids at a slow rate.

soda ash *n*: sodium carbonate (Na_2CO_3).

sodium acid pyrophosphate (SAPP) *n*: a thinner used in combination with barite, caustic soda, and fresh water to form a plug and seal off a zone of lost circulation.

sodium bicarbonate *n*: the half-neutralized sodium salt of carbonic acid, $NaHCO_2$, used extensively for treating cement contamination and occasionally other calcium contamination in drilling fluids.

sodium carbonate *n*: Na_2CO_3, used extensively for treating various types of calcium contamination. Also called soda ash.

sodium hydroxide *n*: see *caustic soda.*

sodium polyacrylate *n*: a synthetic high-molecular-weight polymer of acrylonitrile used primarily as a fluid loss-control agent.

sodium silicate muds *n pl*: special class of inhibited chemical muds using as their bases sodium silicate, salt, water, and clay.

soft water *n*: water that does not contain dissolved calcium and magnesium salts.

solids *n pl*: see *mud solids.*

solids concentration *n*: total amount of solids in a drilling fluid as determined by distillation. Includes both the dissolved and the suspended or undissolved solids.

solids content *n*: see *solids concentration.*

sour corrosion *n*: embrittlement and subsequent wearing away of metal caused by contact of the metal with hydrogen sulfide.

spall *v*: to break off in chips or scales, as on a plain, or journal, bearing.

specific gravity *n*: the ratio of the density, or weight, of a substance to the density of a reference substance. For liquids and solids, the reference substance is water.

sp gr *abbr*: specific gravity.

spot *v*: to pump a designated quantity of a substance (such as acid or cement) into a specific interval in the well. For example, 10 barrels (1,590 litres) of diesel oil may be spotted around an area in the hole in which drill collars are stuck against the wall of the hole in an effort to free the collars.

spud in *v*: to begin drilling; to start the hole.

spud mud *n*: the fluid used when drilling starts at the surface, often a thick bentonite-lime slurry.

static pressure *n*: the pressure exerted by a fluid upon a surface that is at rest in relation to the fluid.

still *n*: instrument used to distill oil, water, and other volatile materials in a mud to determine oil, water, and total solids contents in volume-percent. Also called a retort.

surfactant *n*: a soluble compound that concentrates on the surface boundary between two substances such as oil and water and reduces the surface tension between the substances. The use of surfactants permits the thorough surface contact or mixing of substances that ordinarily remain separate. Surfactants are used in the petroleum industry as additives to drilling mud and to water during chemical flooding. Also called wetting agent, emulsifier.

surfactant mud *n*: a drilling mud prepared by adding a surfactant to a water-base mud to change the colloidal state of the clay from that of complete dispersion to one of controlled flocculation. Such muds were originally designed for use in deep, high-temperature wells, but their many advantages (high chemical and thermal stability, minimum swelling effect on clay-bearing zones, lower plastic viscosity, and so on) extend their applicability.

suspending agent *n*: an additive used to hold the fine clay and silt particles that sometimes remain after an acidizing treatment in suspension; i.e., it keeps them from settling out of the spent acid until it is circulated out.

suspension *n*: a mixture of small nonsettling particles of solid material within a gaseous or liquid medium.

SW *abbr*: salt water; used in drilling reports.

swabbing effect *n*: a phenomenon characterized by formation fluids being pulled or swabbed into the wellbore when the drill stem and bit are pulled up the wellbore fast enough to reduce the hydrostatic pressure of the mud below the bit. If enough formation fluid is swabbed into the hole, a kick can result.

sx *abbr*: sacks; used in drilling and mud reports.

synthetic-based mud *n*: a drilling fluid containing man-made chemicals that emulate natural oil. Natural oil-base muds require that the cuttings made by the bit be specially handled to prevent damage to the environment; synthetic muds were therefore developed to replace oil-base muds in environmentally sensitive areas. For this reason, synthetic-based muds are sometimes termed pseudo-oil-base mud. See *oil-base mud.*

T

t *sym*: tonne.

temperature *n*: a measure of heat or the absence of heat, expressed in degrees Fahrenheit or Celsius. The latter is the standard used in countries on the metric system.

temperature gradient *n*: 1. the rate of change of temperature with displacement in a given direction. 2. the increase in temperature of a well as its depth increases.

ten-minute gel strength *n*: the measured 10-minute gel strength of a fluid is the maximum reading (deflection) taken from a direct-reading viscometer after the fluid has been quiescent for 10 minutes. The reading is reported in pounds/100 square feet. See *gel strength*.

thin *v*: to add a substance such as water or a chemical to drilling mud to reduce its viscosity.

thinning agent *n*: a special chemical or combination of chemicals that, when added to a drilling mud, reduces its viscosity.

thixotropy *n*: the property exhibited by a fluid that is in a liquid state when flowing and in a semisolid, gelled state when at rest. Most drilling fluids must be thixotropic so that cuttings will remain in suspension when circulation is stopped.

titration *n*: a process of chemical analysis by which drops of a standard solution are added to another solution or substance to obtain a desired response—color change, precipitation, or conductivity change—for measurement and evaluation.

treat *v*: to subject a substance to a process or to a chemical to improve its quality or to remove a contaminant.

trip *n*: the operation of hoisting the drill stem from and returning it to the wellbore. *v*: shortened form of "make a trip."

trip in *v*: to go in the hole.

trip margin *n*: the small amount of additional mud weight carried over that needed to balance formation pressure to overcome the pressure-reduction effects caused by swabbing when a trip out of the hole is made.

trip out *v*: to come out of the hole.

trip sheet *n*: a record of the measured drilling fluid that displaces the drill string as one pulls the pipe out of the hole or runs the pipe into the hole. See *trip tank.*

trip tank *n*: a small mud tank with a capacity of 10 to 15 barrels (1,590 to 2,385 litres), usually with 1-barrel or ½-barrel (159-litre or 79.5-litre) divisions, used to ascertain the amount of mud necessary to keep the wellbore full with the exact amount of mud that is displaced by drill pipe. When the bit comes out of the hole, a volume of mud equal to that which the drill pipe occupied while in the hole must be pumped into the hole to replace the pipe. When the bit goes back in the hole, the drill pipe displaces a certain amount of mud, and a trip tank can be used again to keep track of this volume.

U

unconsolidated shales *n pl*: compacted beds of clays and silts that are soft.

underbalanced *adj*: of or relating to a condition in which pressure in the wellbore is less than the pressure in the formation.

underbalanced drilling *v*: to carry on drilling operations with a mud whose density is such that it exerts less pressure on bottom than the pressure in the formation while maintaining a seal (usually with a rotating head) to prevent the well fluids from blowing out under the rig. Drilling under pressure is advantageous in that the rate of penetration is relatively fast; however, the technique requires extreme caution.

V

velocity *n*: 1. speed. 2. the timed rate of linear motion.

V-G meter *n*: see *Fann V-G™ meter*; *direct-indicating viscometer.*

viscometer *n*: a device used to determine the viscosity of a substance. Also called a viscosimeter.

viscosity *n*: a measure of the resistance of a fluid to flow. Resistance is brought about by the internal friction resulting from the combined effects of cohesion and adhesion. The viscosity of petroleum products is commonly expressed in terms of the time required for a specific volume of the liquid to flow through a capillary tube of a specific size at a given temperature.

viscous *adj*: having a high resistance to flow.

volume *n*: the amount of a substance that occupies a particular space.

W

wall-building ability *n*: the ability of a drilling mud to form a filter cake.

wall cake *n*: also called filter cake or mud cake. See *filter cake.*

washout *n*: 1. excessive wellbore enlargement caused by solvent and erosional action of the drilling fluid. 2. a fluid-cut opening caused by fluid leakage.

water-back *v*: 1. to reduce the weight or density of a drilling mud by adding water. 2. to reduce the solids content of a mud by adding water.

water-base mud *n*: a drilling mud in which the continuous phase is water. In water-base muds, any additives are dispersed in the water. Compare *oil-base mud.*

water loss *n*: see *fluid loss.*

weight *n*: see *mud weight.*

weighting material *n*: a material that has a high specific gravity and is used to increase the density of drilling fluids or cement slurries.

weight up *v*: to increase the weight or density of drilling fluid by adding weighting material.

wetting *n*: the adhesion of a liquid to the surface of a solid.

wetting agent *n*: see *surfactant.*

whole mud *n*: all the components of the drilling mud, including reactive and nonreactive solids and the liquid phase (filtrate).

X

X-C polymer *n*: a polymer produced by the action of a particular strain of bacteria on carbohydrates, used for increasing apparent viscosity and yield point with moderately good filtration control. Also called biopolymer.

Y

yield *n*: the number of barrels of a liquid slurry of a given viscosity that can be made from a ton of clay. Clays are often classified as either high- or low-yield. A ton of high-yield clay yields more slurry of a given viscosity than a low-yield clay. Also called clay yield.

yield point *n*: 1. the maximum stress that a solid can withstand without undergoing permanent deformation either by plastic flow or by rupture. 2. in drilling mud engineering, another term for yield value.

yield value *n*: the resistance to initial flow, or the stress required to start fluid movement. This resistance is caused by electrical charges located on or near the surfaces of the particles. The values of the yield point and thixotropy, respectively, are measurements of the same fluid properties under dynamic and static states. The Bingham yield value, reported in pounds/100 square feet, is determined from a direct-indicating viscometer by subtracting the plastic viscosity from the 300-rpm reading. Also called yield point.

Z

zinc chloride ($ZnCl_2$) *n*: a very soluble salt used to increase the density of water to points more than double that of plain water. Normally added to a system first saturated with calcium chloride.

zone of lost circulation *n*: a formation that contains holes or cracks large enough to allow cement to flow into the formation instead of up along the annulus outside of the casing.

Review Questions

LESSONS IN ROTARY DRILLING

Unit II, Lesson 2: Drilling Fluids

Multiple Choice

Pick the *best* answer from the choices and place the letter of that answer in the blank provided.

_________ 1. All of the following are functions of drilling fluids, *except*—
- a. transmits hydraulic power to the bit
- b. prevents entry of formation fluids into the well
- c. keeps the drill bit and drill stem hot
- d. cleans the bottom of the hole

_________ 2. Assuming that a rig is using drilling mud as the circulating fluid and the driller stops circulating to allow crew members to make a connection, which of the following should occur?
- a. The cuttings fall immediately to the bottom of the hole.
- b. The drill collars become lighter in weight.
- c. The jets of mud leaving the bit increase in velocity.
- d. The cuttings remain suspended in the hole.

_________ 3. Normally, the main purpose of the high-velocity streams of drilling fluid that jet out of a conventional bit is to—
- a. move cuttings away from the formation face as quickly as the bit cuts them.
- b. gouge out large portions of the bottom of the hole.
- c. ensure that the cuttings stay in place on the bottom of the hole.
- d. put a layer of solids on the bottom of the hole.

_________ 4. All of the following are functions of drilling fluids, *except*—
- a. reveals the presence of oil, gas, or water that may enter the circulating fluid from a formation being drilled
- b. forces formation rocks away from the wellbore to enlarge it sufficiently for the bit to drill properly
- c. transports bit cuttings to the surface
- d. reveals information about formations by means of the cuttings brought to the surface

_________ 5. Viscosity is a measure of a fluid's—
- a. resistance to flow.
- b. ability to freeze.
- c. resistance to temperature.
- d. melting point.

_________ 6. Slip velocity is a measure of the rate at which—
  a. the bit drills.
  b. cuttings settle.
  c. crew members set the slips.
  d. drilling mud flows.

_________ 7. Gel strength is a measure of a drilling mud's ability to—
  a. form wall cake.
  b. lift cuttings from the bottom of the hole.
  c. prevent the entry of formation fluids into the wellbore.
  d. suspend cuttings in the hole when circulation stops.

_________ 8. Wall cake forms when—
  a. pressure forces mud solids into the formation, leaving behind mud liquids, which are plastered to the borehole wall.
  b. formation pressure forces solid particles of the formation into the wellbore.
  c. plaster of paris is added to the mud.
  d. pressure forces mud liquids into the formation, leaving behind mud solids, which are plastered to the borehole wall.

_________ 9. Wall cake—
  a. stabilizes the hole.
  b. prevents the hole from caving in.
  c. minimizes fluid loss.
  d. all of the above

_________ 10. Which of the following muds exerts the most pressure on the bottom of the hole?
  a. 10 ppg (1,198 kg/m^3)
  b. 10.2 ppg (1,222 kg/m^3)
  c. 12.0 ppg (1,438 kg/m^3)
  d. 14.7 ppg (1,761 kg/m^3)

True or False

Put a T for *true* and an F for *false* in the blank next to each statement.

_________ 11. Because air or gas drilling fluids can make it possible to drill a hole much faster than using liquid drilling fluids, the most common drilling fluids used in drilling are air or gas.

_________ 12. Drilling mud consists of three phases: a liquid phase and two solid phases.

_________ 13. Oil is never used in making up a drilling mud; only water is used.

_________ 14. Barite is a reactive solid added to drilling mud to increase its weight, or density.

_________ 15. The solids content of a drilling mud refers to solids purposely added to the mud to give it desired properties as well as drilled solids.

Matching

Write the letter of the correct definition in the blank next to each term.

Terms

________ 16. blowout

________ 17. kick

________ 18. underbalanced drilling

________ 19. lost circulation

________ 20. barite

________ 21. desander and desilter

________ 22. gas cutting

________ 23. viscosity

________ 24. gel strength

________ 25. yield point

________ 26. bentonite

________ 27. polymers

________ 28. dispersants

________ 29. fluid loss

________ 30. pH

Definitions

a. The liquid part of the mud (oil or water) that enters a porous and permeable formation opposite the wellbore.

b. Hydrates in water to form a gel.

c. The unexpected flow of formation fluids into the wellbore.

d. Form powerful colloids and gels and dissolve in water.

e. Devices that remove drilled solids from the mud.

f. A mineral with a specific gravity of 4.2 to 4.35 and used to increase the density of drilling mud.

g. A fluid's resistance to flow.

h. A measure of a mud's ability to suspend cuttings.

i. The uncontrolled flow of formation fluids into the atmosphere or into another formation exposed to the wellbore.

j. The alkalinity or acidity of a fluid.

k. The loss of whole quantities of mud to a formation; usually occurs because the mud weight is too heavy for the formation being drilled and fractures it.

l. A measure of force required to make a mud flow.

m. Deliberately keeping the hydrostatic pressure below formation pressure to increase the drilling rate.
n. The entry of small amounts of gas from a drilled formation.
o. Reduce viscosity and gel strength.

Multiple Choice

Pick the *best* answer from the choices and place the letter of that answer in the blank provided.

________ 31. The most commonly used drilling fluids are—
- a. oil-base muds.
- b. air or gas.
- c. water-base muds.
- d. oil-emulsion muds.

________ 32. A natural mud is a mud that—
- a. contains no water.
- b. is made up of many chemicals besides water.
- c. contains water and formation clays.
- d. none of the above

________ 33. Funnel viscosity is a measure of—
- a. how long it takes a quart or litre of mud to flow through a Marsh funnel.
- b. the temperature of a mud.
- c. the amount of solids in the mud.
- d. the thickness of a plastic funnel.

________ 34. Which of the following chemicals are *not* used to viscosify (increase the viscosity of) a mud?
- a. bentonite
- b. barite
- c. attapulgite
- d. polymers

________ 35. Which of the following substances are not used to increase the density of a mud?
- a. barite
- b. defoamers
- c. hematite
- d. galena

________ 36. Lost circulation materials include—
- a. dissolved salts.
- b. lignosulfonates.
- c. phosphates.
- d. none of the above

________ 37. A low-solids mud is a mud in which—
a. drilled solids are kept to a maximum.
b. sand and silt in the mud are dispersed.
c. the amount of drilled solids is controlled.
d. barite is recovered by a centrifuge.

________ 38. Shear thinning is—
a. adding water to a mud to make it less viscous.
b. an increase in viscosity that occurs when pressure is applied to make a mud flow.
c. a decrease in viscosity that occurs when pressure is applied to make a mud flow.
d. adding polymers to mud to make it less viscous.

________ 39. Aerated mud is mud—
a. into which air or gas is injected.
b. that has formation gas in it.
c. to which oil has been added.
d. none of the above

________ 40. An oil-emulsion mud is—
a. an oil-base mud to which water has been added.
b. a water-base mud to which oil has been added.
c. an oil-base mud that contains no water.
d. an oil-base mud that contains mostly oil.

True or False

Put a T for *true* or an F for *false* in the blank next to each statement.

________ 41. Oil muds are often employed because they are cheaper than water-base muds.

________ 42. Oil muds do not lubricate the bit and drill stem as well as water-base muds.

________ 43. An oil-base mud contains less than 5 percent water.

________ 44. An invert-emulsion mud is a water-base mud to which some oil has been added.

________ 45. A synthetic mud is an oil mud that contains a biodegradable oil or an oil that is less toxic than diesel or crude oil.

________ 46. Penetration rates are usually higher with air or gas than with mud.

________ 47. When drilling with air or gas, the entry of water into the wellbore from a formation can be a problem.

________ 48. Adding a foaming agent to air or gas can make it possible to continue drilling even though water is entering the wellbore.

Matching

Write the letter of the correct definition in the blank next to each term.

Terms

_________ 49. connate water

_________ 50. sloughing shale

_________ 51. overburden pressure

_________ 52. differential pressure sticking

_________ 53. lost circulation

_________ 54. oil mud

_________ 55. outcropping rock

_________ 56. 0.052

_________ 57. circulating pressure losses

_________ 58. swabbing

_________ 59. kick prevention

_________ 60. drilling break

_________ 61. trip tank

_________ 62. shutting in a well

_________ 63. true vertical depth

_________ 64. hydrogen sulfide (H_2S)

_________ 65. caustic soda

Definitions

a. A formation that breaks off into the hole, causing problems.
b. A sudden increase in the rate of penetration.
c. A highly toxic (poisonous) gas that can kill if proper steps are not taken.
d. The closing of all valves and blowout preventers to prevent the further entry of formation fluids into the wellbore after a kick occurs.
e. All or part of the circulating mud fails to return to the surface.
f. Pressure lost because of a fluid's rubbing against various components as it circulates.
g. The pulling of formation fluids into the wellbore as the drill string is removed from the wellbore.
h. A phenomenon that occurs when the drill collar string becomes embedded in thick wall cake because of pressure in the borehole being considerably higher than pressure in the formation.

i. Keeping the hole full of mud of the correct weight, or density.
j. A mud able to tolerate temperatures up to 450°F (232°C).
k. A relatively small tank with a finely graduated gauge that shows how much mud has been pumped into the hole to replace the drill string.
l. A strong alkaline substance that can cause serious burns to the skin if not handled properly.
m. Relieves pressure placed on a fluid-filled formation by overburden.
n. The depth of a well measured in a straight (vertical) line from the surface to the bottom of the well.
o. Pressure that overlying rocks exert on a particular formation.
p. Water that existed in the formation when it was formed in the ancient past.
q. A mathematical constant used to calculate pressure gradient.

True or False

Put a T for *true* or an F for *false* in the blank next to each statement.

_________ 66. When performing field tests on drilling mud, it is not necessary to duplicate downhole conditions to get an accurate test.

_________ 67. Air in mud can give erroneous test results.

_________ 68. A mud balance is the usual device on the rig for measuring a mud's density.

_________ 69. Two methods of measuring a mud's viscosity include a Smith funnel and a direct-indicating viscometer.

_________ 70. A direct-indicating viscometer measures gel strength.

_________ 71. A mud still is used to measure solids content.

_________ 72. Chemically treated paper strips are used to measure a mud's pH.

_________ 73. Clay percentage in a mud sample can be measured by using a dye called methylene blue.

_________ 74. Titration is a method for analyzing filtrate alkalinity.

_________ 75. Temperature of an oil-base mud is not important when measuring its funnel viscosity.

Answers to Review Questions

LESSONS IN ROTARY DRILLING

Unit II, Lesson 2: Drilling Fluids

1. c
2. b
3. a
4. d
5. a
6. b
7. d
8. a
9. d
10. d
11. F
12. T
13. F
14. F
15. T
16. i
17. c
18. m
19. k
20. f
21. e
22. n
23. g
24. h
25. l
26. b
27. d
28. o
29. a
30. j
31. c
32. c
33. a
34. b
35. b
36. d
37. c
38. c
39. a
40. b
41. F
42. F
43. T
44. F
45. T
46. T
47. T
48. T
49. p
50. a
51. o
52. h
53. e
54. j
55. m
56. q
57. f
58. g
59. i
60. b

61. k
62. d
63. n
64. c
65. l
66. F
67. T
68. T
69. F
70. T
71. T
72. T
73. T
74. T
75. F